NOVA EXCERPTS

EARTH SCIENCES

NOVA EXCERPTS

Additional books and e-books in this series can be found on Nova's website under the Series tab.

NOVA EXCERPTS

EARTH SCIENCES

EDITOR

nova science publishers
New York

NOTICE TO THE READER

Library of Congress Cataloging-in-Publication Data

ISBN: 978-1-53619-038-0

Published by Nova Science Publishers, Inc. † New York

Contents

PREFACE

Nova Excerpts: Earth Sciences is a compilation of Abstracts and Introductions from a wide variety of Earth Science books Nova has published over the years. Each excerpt includes the title of the original book where the original content was published should readers be interested in learning more. *Nova Excerpts* is published simultaneously in both print and digital formats.

In: Nova Excerpts: Earth Sciences
Editor: Dominik Lucas
ISBN: 978-1-53619-038-0

Chapter 1

EXPERIMENTAL STEPS FOR CALIBRATING A RADON METER FOR USE IN ENVIRONMENTAL STUDIES*

Eder Queiroz Barbosa and Daniel Marcos Bonotto†
Departamento de Petrologia e Metalogenia,
UNESP-São Paulo State, Rio Claro, São Paulo, Brazil

ABSTRACT

Potential health hazards from radon in consuming water have been considered worldwide, with many countries adopting the guideline activity concentration of 100 Bq/L for drinking water quality as proposed in the past by WHO. Radon is quickly absorbed by inhalation, through mucous membranes, rather than by bathing. In general, such recommendation was applied to routine operational conditions of water supply systems despite special attention was given if ground waters are utilized for public water supplies as their ^{222}Rn levels are commonly 10-100 times higher than the values in surface waters (often less than 3.7 Bq/L). Typically, about 70% to 90% of the radon gas dissolved in water escapes into the atmosphere,

---

* The full version of this chapter is published in *Radon: Detection, Exposure and Control*, edited by Ramesh Chandra Tiwari, published by Nova Science Publishers, Inc., 2020.

† Corresponding Author's Email: danielmarcosbonotto@gmail.com.

implying on the requirement of water possessing 10,000 Bq/L of dissolved ^{222}Rn for each 1Bq/L in the air. Thus, an indoor environment with water containing 1,000 Bq/L of dissolved ^{222}Rn will contribute to the radon release into the air of about 100 Bq/m^3. All these aspects indicate that environmental studies focusing the presence of radon in the atmosphere and waters require properly calibrated equipments based on standardized procedures as the data generated will be compared with guideline reference values for health risks at different national and international levels. A wide range of systems have been developed for the radon measurements elsewhere and this chapter describes a set of laboratorial experiments held in order to reach this purpose for the Pylon AB6A that is a equipment employing a scintillation detector consisting of a Lucas cell coupled to a photomultiplier tube.

Keywords: lucas cell, scintillation technique, pylon radon meter, continuous analysis method, quasi analysis method

INTRODUCTION

Radon (^{222}Rn, half-life 3.8 days) is a naturally occurring volatile noble gas formed from the ^{238}U radioactive decay series, according to the following sequence: ^{238}U (4.49 Ga, α) $\rightarrow ^{234}Th$ (24.1 d, β^-) $\rightarrow ^{234}Pa$ (1.18 min, β^-) $\rightarrow ^{234}U$ (0.248 Ma, α) $\rightarrow ^{230}Th$ (75.2 ka, α) $\rightarrow ^{226}Ra$ (1622a, α) $\rightarrow ^{222}Rn$ (3.8d, α) $\rightarrow$ ... It is a colorless, odorless, tasteless, chemically inert and radioactive gas produced continuously in rocks and soils through ^{226}Ra α-decay, with some atoms escaping to the surrounding fluid phase, such as groundwater and air. It is subjected to recoil at "birth", with the emanated fraction relatively to that produced in the solid phase being dependent on factors such as total surface area of solids and concentration/distribution of ^{238}U (^{226}Ra) in the minerals. ^{222}Rn decays to stable lead according to the sequence: ^{222}Rn (3.8d, α) $\rightarrow ^{218}Po$ (3.05 min, α) $\rightarrow ^{214}Pb$ (26.8 min, β^-) $\rightarrow ^{214}Bi$ (19.7 min, β^-) $\rightarrow ^{214}Po$ (0.16 ms, α) $\rightarrow ^{210}Pb$ (22.3 a, β^-) $\rightarrow ^{210}Bi$ (5 d, β^-) $\rightarrow ^{210}Po$ (138.4 d, α) $\rightarrow ^{206}Pb$.

High ^{222}Rn concentrations occur in groundwaters in many areas where wells are used for domestic water supply, inclusive in small rural water supplies (Petrie and Cram, 1999). Exposure to radon and its progeny is believed to be associated with increased risks of several kinds of cancer (USEPA, 1999). When radon or its progeny are inhaled, lung cancer accounts for most of the total incremental cancer risk, while ingestion of radon in water is suspected of being associated with increased risk of tumors of several internal organs, primarily the stomach (USEPA, 1999). Inhalation of radon progeny accounts for about 89% of the individual risk associated with domestic water use, with almost 11% resulting from directly ingesting radon in drinking water (USEPA, 1999). Gesell and Prichard (1975, 1980), Partridge et al. (1979), Bruno (1983), Prichard (1987), Nazaroff and Nero (1988) and Nazaroff et al. (1988) also pointed out that although radon in water is not a well-documented health risk, it does contribute to radon in indoor air, which has been established as a health threat. Furthermore, high concentrations of ^{222}Rnin groundwaters also indicate the presence of radon's parent nuclides, ^{238}U and ^{226}Ra, in the water-rock system, which are known health risks when ingested in drinking water (Aieta et al., 1987).

Beyond the importance to develop effective methods for reducing or eliminating radon from homes, the knowledge of the geochemical and hydrogeological mechanisms that control the movement of ^{222}Rn through the groundwater system may be informative of the natural processes related to its high concentration associated with low transmissivity zones (Lawrence et al., 1991), with the uranium content of the source rock, severe chemical weathering, hydrothermal solution, deposition, or extensive fracturing (Nelson et al., 1983), and with variations in stress in rocks associated with seismicity, where in some cases temporal variations have preceded large earthquakes (Shishkevich, 1971; Smith et al., 1976; Wollenberg et al., 1984).

Several field and laboratory studies have been conducted and many models for radon generation have been proposed, most of which are applicable for a single field area or have been developed using microscopic properties. For instance, Nelson et al. (1983) proposed a simple crack model to understand the relation between flow rate and radon concentration, where water flows radially inward from all directions through a thin flat disk into the borehole, to be collected within the central hole; different assumptions for mixing and for entry point locations were performed. Rama and Moore

(1984) suggested that radon diffuses to the atmosphere or water through a network of 100 - 200 Å wide nanopores in rocks and minerals, many of which intersect the grain surface, and give rise to a large internal surface area of the solids (in the range of about 10 m^2 per cm^3), orders of magnitude higher than the external surface calculated from grain size. Krishnaswami and Seidemann (1988) pointed out that the high ^{222}Rn leakage from rocks and minerals responsible for regulating the supply of this gas to the surrounding fluid phases such as air and groundwater is related to ^{226}Ra enrichment in grain boundaries or crevices instead of an extensive network of nanopores, which could increase their total surface area several fold. Semkow (1990) evaluated the mechanism of α-recoil to account for Rn emanation from solids, emphasizing the basic phenomena related to single mineral grains, since soils and rocks could be considered multi-grain aggregates; the author derived formulas for the emanation from cylinders and spheres having non-uniform Ra distribution or variable thickness of Ra-containing material.

Potential health hazards from radon in consuming water have been considered worldwide, with many countries adopting the guideline activity concentration of 100 Bq/L for drinking water quality recommended by WHO (2008). In general, the recommendations apply to routine operational conditions of water supply systems, however, special attention must be given when ground waters are utilized for public water supplies, because ^{222}Rn concentrations in surface waters are often less than 3.7 Bq/L, while in groundwater the ^{222}Rn concentrations commonly are 10-100 times higher (Wanty et al., 1992).

Thus, the Rn characterization in waters is important because it may utilized as an isotopic tracer in hydrologic investigations (Hoehn and von Gunten, 1989) and also in order to assure the drinking water quality, since although radon in water is not a well-documented health risk, it does contribute to radon in indoor air, which has been established as a health threat (Nazaroff and Nero, 1988). A convenient method often used for quantifying radon dissolved in water is based on the emanation procedure (Lucas, 1957), among other reported in the literature. It consists on its removal from the sample, its transfer to a scintillation flask, and its detection by alpha-scintillation counting. Such technique has been widely utilized in environmental studies focusing the presence of dissolved radon and slightly modified by different manufactures. In general, the α-counting is carried out following a delay of about 3 h from the Rn transfer to permit total ingrowth of the short-lived ^{222}Rn daughters, with the flask being placed on the end of a photomultiplier tube covered with a light-tight lid. A counter/timer is then used for recording the total number of pulses, being performed a calibration in order to convert this parameter in a value of activity concentration. Thus, properly calibrated equipments based on standardized procedures are required for the radon data acquisition, as they will be compared with guideline reference values for health risks at different national and international levels (DFPM, 1966; WHO, 2008). This chapter describes a set of laboratorial experiments held in order to reach this purpose for the Pylon AB6A that is an equipment employing a scintillation detector consisting of a Lucas cell coupled to a photomultiplier tube.

In: Nova Excerpts: Earth Sciences
Editor: Dominik Lucas
ISBN: 978-1-53619-038-0

Chapter 2

THE INFLUENCE OF REGIONAL AND GLOBAL GEOPHYSICAL CHARACTERISTICS ON THE VOLUMETRIC ACTIVITY OF RADON IN GORNO-ALTAISK (RUSSIA) *

Alexander V. Shitov †
Geography Faculty, Gorno-Altaisk State University, Gorno-Altaisk, Russia

ABSTRACT

The author makes an assumption about the influence of global (solar and geomagnetic activity, the angular velocity of the earth's rotation, etc.) and local geophysical processes (pressure, seismic activity) on the dynamics of the volume activity of radon. On the example of studying the dynamics of radon volume activity in Gorno-Altaisk (Russia) for the period 2012-2016, a comparative study of these characteristics was made and the influence and relationship of global processes on the dynamics of radon volume activity was shown. The volume activity of radon (VAR) has a common periodization with regional ones: atmospheric pressure, seismic

* The full version of this chapter is published in *Radon: Detection, Exposure and Control*, edited by Ramesh Chandra Tiwari, published by Nova Science Publishers, Inc., 2020.

† Corresponding Author's Email: sav103@yandex.ru.

activity, potential difference and global characteristics: geomagnetic index Kp, angular velocity of the earth's rotation. The VAR has a common periodization with regional ones: atmospheric pressure, seismic activity, potential difference and global characteristics: geomagnetic index Kp, angular velocity of the earth's rotation. Comparison of the studied regional and global characteristics showed that the influence of seismic activity on the VAR is ambiguous.

Keywords: volumetric activity of radon, monitoring, earthquakes, meteorological characteristics, Altai

Introduction

Modern complex study of various geophysical characteristics allows to analyze the dynamics of these parameters, their interdependence, as well as the influence of other global processes on the studied characteristics. A large number of works devoted to the study of the influence of heliogeophysical characteristics on the seismic activity of different regions are known, the interdependence of meteorological parameters and seismic activity is known, in turn, the preparation of large seismic events affects a large complex of geological and geophysical characteristics not only of the region in which an earthquake occurs, but also can (Doda, Stepanov et al., 2013; Aptikaeva, Shitov, 2016).

Our work on the study of the dynamics of the volumetric activity of radon over a long period of time allowed us to consider together with it the dynamics of various regional characteristics, as well as a number of global geophysical processes, and may allow us to show a number of influences and interactions of these processes.

In: Nova Excerpts: Earth Sciences
Editor: Dominik Lucas
ISBN: 978-1-53619-038-0

Chapter 3

FEATURES OF RADON DISTRIBUTION OVER KIMBERLITE PIPES OF THE ARKHANGELSK DIAMONDIFEROUS PROVINCE: POSSIBLE DIRECTIONS FOR THE STUDY OF THE STRUCTURE NEAR-PIPE SPACE AND DEVELOPMENT OF THE EMANOMETRIC METHOD FOR THE SEARCH FOR KIMBERLITES*

Evgeny Yu. Yakovlev*†, *Sergey V. Druzhinin*, *Sergey B. Zykov and Elena N. Zykova
Institute of Geodynamics and Geology, N. Laverov Federal Centre for Integrated Arctic Research (FCIARctic), Arkhangelsk, Russia

* The full version of this chapter is published in Radon*: Detection, Exposure and Control*, edited by Ramesh Chandra Tiwari, published by Nova Science Publishers, Inc., 2020.

† Corresponding Author's Email: evgeny.yakovlev@fciarctic.ru.

Abstract

Field survey of the volume activity of radon in soil air over kimberlite pipes of the Zolotitsky field of the Arkhangelsk diamondiferous province was carried out. In the course of the emanometric survey, it was found that kimberlite pipes are distinguished by the sub-circular anomalies of radon activity, exceeding the background values by 4 to 48 times. The most pronounced radon anomalies are formed in the space near the pipes, in which the development of increased fracturing zones is typical, creating favorable conditions for radon migration towards the surface. The revealed regularities in the variations of radon activity create good prerequisites for the further development of the emanometric method for the search for kimberlite bodies.

Keywords: geology, kimberlite, radon, radioactivity, emanometric methods

Introduction

Since the discovery of the first kimberlite pipe in the Arkhangelsk diamondiferous province (ADP), many traditional geophysical methods have been tested for the purpose of the kimberlite body search, and among them, aeromagnetic methods have shown the best performance, as in many diamondiferous provinces around the world (Kontarovich and Tsyganov, 2000; Tsyganov et al., 2004; Babayants et al., 2006; Kontarovich and Babayants, 2011). However, currently, there is a certain crisis in the application of magnetic prospecting in the territory of the Arkhangelsk province, which manifests itself in a sharp decline in its search efficiency, which currently is less than 0.5% (Stogniy and Korotkov, 2010; Korotkov 2011). The use of a number of other geophysical, geochemical, and mineralogical methods did not demonstrate any sustainable search efficiency. According to some authors, the main negative factors complicating the search for kimberlites in this area are: the powerful cover of overlying deposits; the hidden type of faults controlling the kimberlite bodies' location; the lean occurrence of kimberlites in physical fields; the weak erosional truncation of the pipes; the extremely low content of ferromagnetic minerals in the pipes; and the low content of associate minerals (Androsov et. al., 2004; Antonov et al., 2004; Ignatov et al., 2011; Lapin and Tolstov, 2007; Frolov et al., 2005; Babayants et al., 2015).

In this connection, it seems highly relevant to develop new approaches to the search for kimberlite pipes and their practical implementation.

The published factual data on the diamondiferous provinces of Russia and other countries confirm the efficiency of certain radiometric methods for the search for kimberlite bodies (Mwenifumbo and Kjarsgaard, 1999; Ramadass et al., 2015).

One of such promising radiometric methods is the soil emanation survey, which has proven itself well in geodynamic studies (Korshunov et al., 2012). Emanation survey is traditionally considered a method that allows identifying faults hidden under deposits, but currently permeable for underground gases (Seminsky et al., 2009; Seminsky et al., 2014). In view of the sufficiently long decay time of radon, the emanation method is most effective in mapping the internal structure of faults, determining the width of fracture zones, and evaluating the degree of tectonic activity of faults (Bobrov 2009). Since it is known that kimberlite bodies are controlled by faults, the emanation studies are deemed important and make it possible to expect a positive result.

Similar emanation studies were performed on kimberlite pipes in South and West Botswana, where the prospects of the emanation method in the prediction of kimberlite bodies were demonstrated (Mc Dowall et al., 1991).

As a result of field radon exploration conducted on the territory of the Chetlassky and Volsko-Vymsky uplifts of Middle Timan (Magomedova et al., 2015), it was found that faults are characterized by increased values of the volume activity of radon. The increased activity of radon is also characteristic of the investigated kimberlite pipes: Umbinskaya, Vodorazdelnaya, Sredninskaya (Magomedova et al., 2016).

This article deals with the study of the volume activity of radon over the known kimberlite pipes of the Zolotitsky field of the Arkhangelsk diamondiferous province. The tasks of the research included the identification of qualitative and quantitative characteristics of the radon field and the estimation of spatial variations in the values of the volume activity of radon over kimberlite bodies.

In: Nova Excerpts: Earth Sciences
Editor: Dominik Lucas
ISBN: 978-1-53619-038-0

Chapter 4

STUDY OF FAULT TRACES IN TAIWAN USING SOIL-GAS GEOCHEMISTRY: AN OVERVIEW*

Vivek Walia[1,†], Arvind Kumar[1], Ching-Chou Fu[3], Wei-Li Hong[4] and Tsanyao Frank Yang[2]

[1]National Center for Research on Earthquake Engineering, National Applied Research Laboratories, Taipei, Taiwan
[2]Department of Geosciences, National Taiwan University, Taipei, Taiwan
[3]Institute of Earth Sciences, Academia Sinica, Taipei, Taiwan
[4]Geological Survey of Norway (NGU), Trondheim Norway, Norway

ABSTRACT

Many studies have shown soil gas method as one of the most reliable investigation tools in the research of earthquake precursory signals and fault delineation. Active faults are important for several reasons. Active fault zones usually have a higher permeability than surrounding strata; therefore, provide conduits for the gases to migrate upward to surface from

---

* The full version of this chapter is published in *Radon:* Detection, *Exposure and Control*, edited by Ramesh Chandra Tiwari, published by Nova Science Publishers, Inc., 2020.

† Corresponding Author's Emails: vivekwalia@rediffmail.com, vivekwalia68@gmail.com, walia@ncree.narl.org.tw.

the deep sources. The present research is aimed to find out the relationship between soil gas distribution and tectonic systems in the vicinity of geologic fault zones in Taiwan. Soil-gas surveys have been conducted across the faults, to find out the degassing pattern of this fault system. During the surveys, soil-gas samples were collected along traverses crossing the observed structures. The collected soil-gas sample bags are analyzed for various gas concentrations (^4He, Ar, O_2, CO_2, N_2) and isotopic compositions ($^3He/^4He$, $\delta^{13}C_{CO2}$, and $\delta\ ^{15}N_{N2}$) in order to understand the origin of the soil gases. The data analysis clearly reveals anomalous values along the faults. The spatial congruence of all the gases shows that nitrogen is the most possible carrier gas of helium, whereas carbon dioxide indicates its carrier gas quality for radon in some area. From the spatial distribution of various gas concentrations the trace of faults and neotectonic features can be identified. Moreover, helium isotopic compositions of representative samples illustrated that most samples have soil air components and may be mixed with some crustal component but no significant input of mantle component. Carbon isotopic composition ($\delta^{13}C$) of carbon dioxide in the soil samples have shown the mixing of organic and limestone components. Both helium and carbon isotopic results support the multiple gas sources in the studied area.

Keywords: soil-gas, fault, earthquake, helium, radon, carrier gases

INTRODUCTION

Identifying active fault is one of the most important aspects of tectonics as seismic hazards are commonly associated with it. Faults can be described as weakened zones composed of highly fractured rock materials, gouge and fluids. Active faults are important for several reasons. Sudden displacements on active faults are the usual cause of earthquakes, and the intensity of the earthquake vibrations at a particular place is partly controlled by the distance to the fault. In addition to damage from earthquake shaking, the displacement on the fault sometimes reaches the ground surface and causes rupture and differential movement that may be as much as several meters. Active faults can provide pathways for gas leaks from the solid earth because they usually increase the permeability of soils. Taiwan is located on the boundary between the Eurasian plate and the Philippine Sea plate. As a result of the continuous stress and collision, Taiwan is densely faulted (Lin et al.,

2000). Therefore, one of the major tasks for disaster mitigation is to delineate the distribution of active faults.

Gas geochemistry has been proven to be a reliable and simple technique to apply, at different scales, to many geological scenarios (Annunziatellis et al., 2003; Lewicki et al., 2003; Baubron et al., 2002; De Gregorio et al., 2002; Kumar et al., 2009, 2013a, b; Walia et al., 2009, 2013). The study of spatial distribution of soil gas anomalies at the surface, can give important and interesting information on the origin and processes involving deep and superficial gas species. This information can be applied and studied in different frameworks, for example: seismic zonation, examining, at the surface, anomalous concentrations of deep gas species that upraise throughout preferential pathways (faults and/or fractures). Soil gas distributions can be directly linked to the evolution of the stress regime and gases migrate preferentially through fractured zones, but only along pathways whose permeability has been enhanced by seismic activity and/or through areas of brittle deformation. Soil gas distribution would be affected by surface features such as pedological, biogenic and meteorological factors: these are supposed to have only a subordinate effect on gas leakage (Hinkle, 1994). However, it is possible to properly interpret soil gas anomalies and recognize the influences of surface features studying the association of different gases (having different origin and physical/chemical behavior), collecting a large number of samples during periods of stable meteorological and soil moisture conditions (e.g., during dry season) and using appropriate statistical treatment of data (experimental variograms to investigate the spatial dependency of gas concentrations). Field data can show the usefulness of the soil gas method for detecting, for instance, crustal discontinuities, even when faults are buried or cut non-cohesive clastic rocks which makes surface recognition difficult using traditional field methods (Ciotoli et al., 1998; Duddridge et al., 1991; Durrance and Gregory, 1988). These characteristics as well as the rapidity and the low cost of the soil gas survey, make this method a powerful tool for geological investigation which can significantly contribute to hazard assessment and forecasting, especially when continuous monitoring is performed (Klusman, 1993; Reimer, 1990; King et al., 1996).

Radon is probably the most useful gas and is used frequently for fault mapping (Al-Taminmi and Abumurad, 2001; Guerra and Lombardi, 2001; Fu et al., 2005; Walia et al., 2005a, 2010) and precursory studies (Walia et al., 2005b, 2006; Yang et al., 2005, 2006). Radon (^{222}Rn) is continuously generated from ^{226}Ra within the rock strata as an intermediate decay product of the ^{238}U radio-active series. The short half-life of ^{222}Rn ($t_{1/2} = 3.82$ d) limits its diffusion in soil, so that Rn measured at the ground surface cannot be released from a deep origin, unless there is a driving mechanism other than mere diffusion. Several models have been invoked to explain Rn migration over large distances and it has been established that Rn is transported by underground water or carrier gases, such as CO_2, CH_4, ^4He or N_2 (Etiope and Martinelli, 2002; Yang et al., 2003b). Its rate of migration and its soil gas concentration are controlled by a large number of factors such as the distribution of U in the soil and bedrock, soil porosity and humidity, microcracks, granulation, surface wind, and so on.

Helium (^4He) has been commonly used as a good fault tracer as it is highly mobile, chemically inert, physically stable, highly insoluble in water, and can travel through meters of fractured overburden. Its distribution in soil gas depends on the permeability of the rocks, depth of the originating source and the influence of near surface conditions. It migrates primarily by advection from deep accumulations toward the surface along permeable fault and fracture systems. Due to these reasons, He has a highly diffusive character with a diffusion coefficient ($1.68 cm^2/s$) about 10 times higher than those of N_2, O_2 and CO_2. It diffuses at a rate much higher than Rn. Helium has a low and constant concentration of 5.239 ± 0.004 ppm in air. Due to these characteristics and the deep origin with respect to Rn, He appears to be a powerful pathfinder for crustal discontinuities, faults and fractures (Ciotoli et al., 1998; Fu et al., 2005). Recent work has observed anomalous ^4He concentrations over faults and confirmed the presence of this gas in shallow soils as a deep fault indicator (Fu et al., 2005; Walia et al., 2005a).

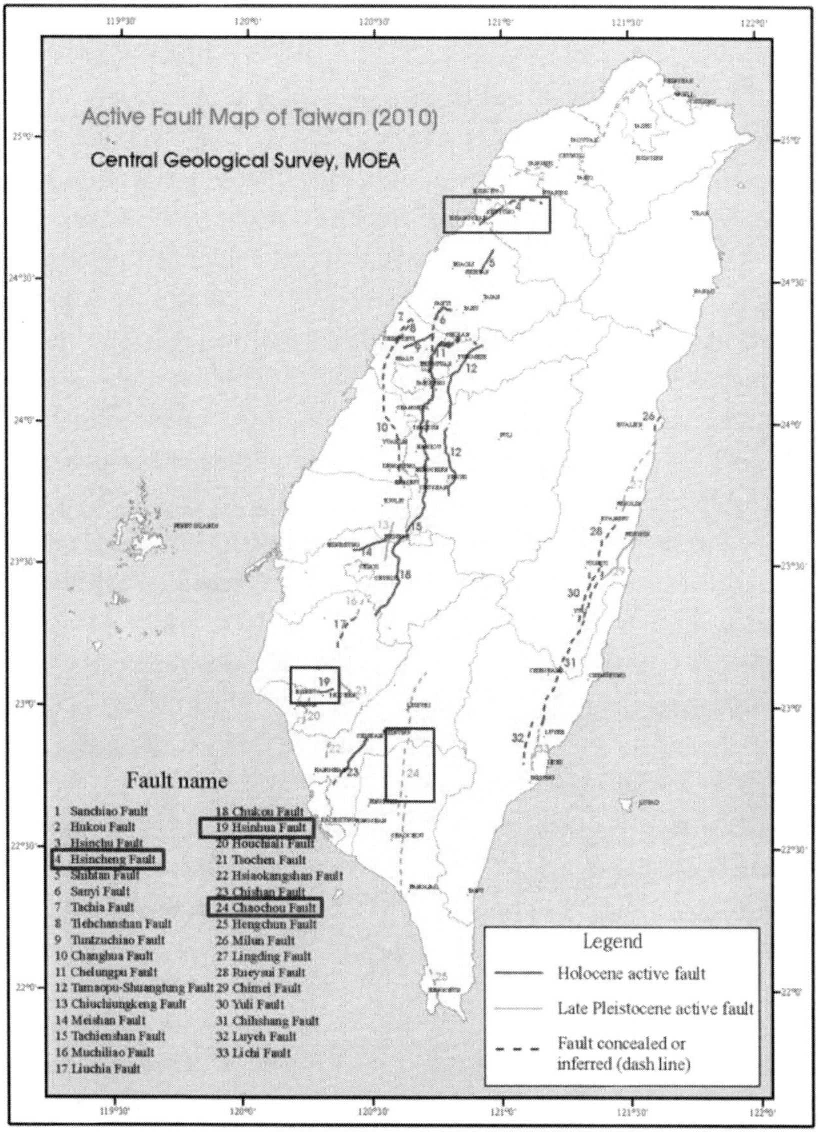

Figure 1. Active fault map of Taiwan and studied faults (marked with red rectangles).

Carbon dioxide is supposed to be the most abundant gas species from hydrothermal to volcanic environments. It is a well defined carrier gas for noble gases such as ^{222}Rn and ^{4}He which are unable to reach the surface due to low mobility and short half-life and too low a concentration, respectively.

Carbon dioxide has several sources: the mantle, metamorphism of carbonate-bearing rocks, decomposition of organic material and surface biological activity (Irwin and Barnes, 1980) and in fault zones CO_2 is a mixture of some of these sources (Fu et al., 2005). High CO_2 fluxes appear to be correlated with both high heat flux areas and limited areas with deep fracturing emitting C originating from the mantle and decarbonation processes, with possible mixing of these two sources. Carbon dioxide discharge indicates areas with high pore pressure at depth, and is therefore used to identify potential seismic regions. Therefore it is used for fault mapping (Irwin and Barnes, 1980; Baubron et al., 1991) as well as for both seismic and volcanic monitoring (Toutain et al., 1992; Rahn et al., 1996; Lan et al., 2007). Nitrogen can also act as a good carrier for noble gases like ^{222}Rn and ^{4}He, and has been detected in soil and spring gases (e.g., Hong et al., 2010). Nitrogen has several sources (e.g., atmospheric, animal and industrial waste, etc.) and can be affected by various physical, chemical and biological processes. The advective movement of ^{222}Rn and ^{4}He must be referenced to carrier gases (viz. CO_2, N_2) that can carry the rare gases. In this chapter, we outline the overview of results obtains from the study of Fault Traces in Taiwan Using Soil-gas Geochemistry method.

The present study is aimed to check the efficiency of soil-gas technique and to determine a possible connection between eventual soil-gas anomalies not only for radon & helium, but also for carrier soil-gases which can be related to some fault systems of Northern and Southern Taiwan (Figure 1). This further can be helpful for continuous monitoring of regional seismic activity and stress build-up of the fault system.

In: Nova Excerpts: Earth Sciences
Editor: Dominik Lucas
ISBN: 978-1-53619-038-0

Chapter 5

RADON EXPOSURE AND MEASUREMENT*

Tayseer Alnaggar†
Department of Physics, College of Women, for Art, Science and Education, Ain Shams University, Cairo, Egypt
Department of Physics, College of Science and Arts, Najran University, Najran, Saudi Arabia

ABSTRACT

Radon is a naturally occurring radioactive gas. It comes from the natural decay of uranium which is found in soil and rocks. Radon travels through soil and enters buildings through cracks and other holes in the foundation. Ultimately, it decays into radioactive isotopes that can become trapped in the lungs at breath. Radon is a known human carcinogen. Extended exposure to elevated radon concentrations causes an increased risk of lung cancer. Like other environmental pollutants. This chapter deal with the measurements of radon by different techniques in soil, rocks, mines, caves, different building materials, sand, water, plants ……., etc. these measurements were introduced by many studies around the world. There are several methods for assessing radon which generally depend on the measurement of their alpha daughters or the gamma emissions of some

* The full version of this chapter is published in *Radon: Detection, Exposure and Control*, edited by Ramesh Chandra Tiwari, published by Nova Science Publishers, Inc., 2020.

† Corresponding Author's Email: ty_sab_sy@yahoo.com.

In: Nova Excerpts: Earth Sciences
Editor: Dominik Lucas
ISBN: 978-1-53619-038-0

Chapter 6

ESTIMATION OF RADON ACTIVITY IN SOIL AND EXHALATION OF RADON FROM SOIL IN BAKRESWAR-TANTLOI GEOTHERMAL REGION OF WEST BENGAL AND JHARKHAND, INDIA*

Saheli Chowdhury[1,2,†], Chiranjib Barman[3], Argha Deb[1,2], Sibaji Raha[4], Debasis Ghose[4] and Bikash Sinha[5]

[1]School of Studies in Environmental Radiation and Archaeological Sciences, Jadavpur University, Kolkata, West Bengal, India
[2]Department of Physics, Jadavpur University, Kolkata, West Bengal, India
[3]Department of Physics, Sidho- Kanho- Birsha University, Purulia, West Bengal, India
[4]Centre for Astroparticle Physics and Space Science, Bose Institute, Kolkata, West Bengal, India
[5]Variable Energy Cyclotron Centre, Kolkata, West Bengal, India

* The full version of this chapter is published in *Radon: Detection, Exposure and Control*, edited by Ramesh Chandra Tiwari, published by Nova Science Publishers, Inc., 2020.

† Corresponding Author's Emails: sahelichowdhury2492@gmail.com, sahelichowdhury.rs@jadavpuruniversity.in.

ABSTRACT

The radioactive gas radon is the predominant terrestrial ionizing radiation that leads to natural radiation exposure. It is ubiquitous in the environment, and its occurrence in nature depends on a number of factors, such as local geological features, soil porosity, grain size, soil permeability, and presence of compounds containing radioactive elements like uranium and radium in soil, apart from local meteorological parameters that control the release of radon into the atmosphere. Inhalation of air containing excess radon is considered harmful on account of radiological hazard posed by its short-lived chemically reactive radioactive decay products. The present work deals with radon estimation in top-soil in Bakreswar-Tantloi Geothermal Region of the states of West Bengal and Jharkhand, India. This geothermal zone lies in the highly faulted Pre-Cambrian Chotanagpur Plateau Region of eastern India. As natural radiation exposure to humans in a region is determined by the amount of radon gas exhaled from the soil into the atmosphere, the exhalation rates of radon from soil have been determined at several locations in this region, and the inhalation doses received by an adult due to such exhalation rates have been estimated.

Keywords: radioactivity, radon, soil radon exhalation, soil gas, geothermal region

INTRODUCTION

Radon is a radioactive noble gas produced by alpha decay of radium in the natural radioactive decay chains in the earth's crust. Radon gas released from soil and water into the atmosphere contributes about 50-70% of the total ionizing radiation to which human beings are exposed (UNSCEAR 2000). Radon gas is a known carcinogen, and exposure to environmental radon is the second leading cause of lung cancer (USEPA 1999). Therefore, study of environmental radon at a place assumes great significance because of the possibility of radiation over-exposure. Of the 39 known isotopes of radon, ^{222}Rn, produced by α-decay of ^{226}Ra, is most abundant in the natural environment due to its comparatively longer half-life of 3.82 days. Although the production of radon through decay of ^{226}Ra ($T_{1/2}$ – 1602 years) could be large, only a small fraction of ^{222}Rn produced in the crust enters into the air-filled pore spaces in soil by diffusion or by alpha recoil. The emanated radon can migrate away from its site of production, either by advection along joints, faults and intergranular pathways in the crust, or being carried by carrier gases like carbon dioxide, methane, etc. or by groundwater circulating deep inside the crust, and can eventually reach the surface layers of soil (Nazaroff et al. 1988, Rogers and Nielson 1991, Nazaroff 1992). The mean distance that ^{222}Rn may travel between its generation and its decay depends upon the property of the soil where it is generated. In particular, the diffusion length depends on the soil diffusion coefficient, which in turn is related to permeability, porosity, grain size and soil water content (Durrani et al. 1997, Rogers and Nielson 1991, Kumar et al. 2011). A part of the emanated radon is exhaled into the atmosphere before decay.

Radon activity in soil, ground water and water bodies, and radon exhalation from the earth surface and different materials have been studied globally for more than six decades by various researchers (Moses et al. 1960, Pearson and Moses 1966, Schweikani et al. 1995, Sohrabi 1998, Kojima and Nagano 1999, Riley et al. 1999, Walia et al. 2005, Espinosa et al. 2008, Ashok et al. 2011, Elzain et al. 2016, Kulali et al. 2017, Sun et al. 2017, Sahoo et al. 2018). It has been reported that the process of emanation, migration and exhalation of radon in the earth's crust is very complex; influenced by various geophysical and geochemical factors including local geology such as presence of faults and shears, soil composition and structure like grain size, porosity and permeability, average uranium concentration in rocks and sediments, and prevalent atmospheric and weather conditions including air temperature, atmospheric pressure, wind flow and precipitation. Also, the release of radon gas from soil as well as from hydrothermal systems has been studied worldwide for about half a century as an important precursor of earthquakes (Tanner 1964, King 1980, Thomas 1988, Hartmann and Levy 2005, Cicerone et al. 2009, Petraki et al. 2015, Barman et al. 2016, Chowdhury et al. 2017). Moreover, radon gas has been utilised as a tracer of geothermal energy reserves and deposits of helium and radioactive substances like uranium and thorium (López et al. 1987, Balcazar et al. 1993, 2010, Khattak et al. 2011).

The exhalation of ^{222}Rn from soil is a major factor affecting human exposure to natural radioactivity in a region. Although radon accumulation due to depleted indoor air circulation in dwellings may cause greater exposure, it is the release of ^{222}Rn from soil that leads to high indoor radon levels in many regions (Nazaroff et al. 1988, Nielson et al. 1994). While radon concentration in groundwater is greater than that in soil gas, the relatively low solubility of radon in water leads to its escape into the surrounding atmosphere. Exhalation of radon into air from soil occurs mostly by molecular diffusion or convection. Exhalation rate is measured as the average activity from a given surface per unit area per unit time. In fact, it is this radon exhaled into the atmosphere that is inhaled by people. Thus, in principle, measurement of ^{222}Rn activity in soil and exhalation of radon gas from soil into atmosphere can be useful to identify potential regions with threat of high natural radiation exposure.

The exhalation of ^{222}Rn from soil is a major factor affecting human exposure to natural radioactivity in a region. Although radon accumulation due to depleted indoor air circulation in dwellings may cause greater exposure, it is the release of ^{222}Rn from soil that leads to high indoor radon levels in many regions (Nazaroff et al. 1988, Nielson et al. 1994). While radon concentration in groundwater is greater than that in soil gas, the relatively low solubility of radon in water leads to its escape into the surrounding atmosphere. Exhalation of radon into air from soil occurs mostly by molecular diffusion or convection. Exhalation rate is measured as the average activity from a given surface per unit area per unit time. In fact, it is this radon exhaled into the atmosphere that is inhaled by people. Thus, in principle, measurement of ^{222}Rn activity in soil and exhalation of radon gas from soil into atmosphere can be useful to identify potential regions with threat of high natural radiation exposure.

^{222}Rn exhalation from soils, rocks, building materials, etc. has been studied extensively throughout the world due to the negative impact of radon on human health (Anastasiou et al. 2003, Nikolopoulos and Louizi 2008, Martín Sánchez et al. 2012). But in most radon exhalation studies, soil or rock samples were collected from sites and exhalation rates were measured in laboratories (Ramola and Choubey 2003; Faheem 2008; Gupta et al. 2010;

Modzelewska et al. 2014; Harb et al. 2015; Elzain et al. 2016). Hence, such studies do not take into account the geological and meteorological factors affecting the exhalation rates at the sites or their daily and/or seasonal variations. Some researchers have performed short-term and long-term in-situ measurements of radon exhalation from soil to study possible influences of local geological and meteorological parameters (Kraner 1964, Pearson and Moses 1966, Fleischer *et al.* 1979, Kojima and Nagano 1999, Ashok *et al.* 2011, Zhang *et al.* 2015, Kandari *et al.* 2016).

In the present work, ^{222}Rn activity in soil has been estimated at 119 locations in Bakreswar-Tantloi Geothermal Region of the states of West Bengal and Jharkhand, India. Also, soil radon exhalation rates have been determined in 28 locations all over the region of interest because the extent of human exposure to natural radioactivity depends heavily on soil radon exhalation rate. Finally, radon exhalation doses experienced by the local people due to such radon exhalation rates from soil have been estimated to form an idea about human exposure to radioactivity in the region. High environmental radioactivity is expected in this region because the bedrock is composed of crystalline igneous and metamorphic rocks from Rajmahal volcanism of the early Cretaceous period. Such rocks are generally highly mineralised with elevated concentrations of minerals containing radioactive elements including uranium. Moreover, this being a heavily faulted region, radon gas generated from decay of radioactive minerals in the bedrock can find easy passages to escape from the parent rocks and reach the upper layers of the terrestrial crust. It has been already observed that groundwater of this region contains large concentration of radon (Ghose *et al.* 2002, Majumdar *et al.* 2010, Naskar *et al.* 2017). Therefore, Bakreswar-Tantloi Geothermal Region is an area of interest for study of environmental radioactivity.

In: Nova Excerpts: Earth Sciences
Editor: Dominik Lucas
ISBN: 978-1-53619-038-0

Chapter 7

RADON BASED BIOMETRIC CHARACTERIZATION FOR PERSON IDENTIFICATION*

Jitendra P. Chaudhari*† and Hiren K. Mewada
Charotar Universiy of Science and Technology, Changa, Anand, Gujarat, India

ABSTRACT

This chapter exclusively focuses on the use of Radon Transform for biometric based person authentication. The Radon is suitable for line parameter extraction. Knuckle print, palmprint, face, iris and hand geometry posse directive line structure (i.e., lines and wrinkles) providing an unique characteristic of an individual. Hence Radon is most suitable for such biometric traits. In addition, the features extracted by the Radon is translation invariant and rotation invariant.

The chapter presents the experimentation of the Radon on fingerprint and palmprint. The first section will describe the basic unimodal and bimodal biometric system. Second section will present region of interest

* The full version of this chapter is published in *Radon: Detection, Exposure and Control*, edited by Ramesh Chandra Tiwari, published by Nova Science Publishers, Inc., 2020.

† Corresponding Author's Email: jitendrachaudhari.ec@charusat.ac.in.

extraction followed by the biometrics feature extraction method using Radon in section three. Fourth section will discuss feature formation using of histogram of Radon transform. The models explained are applied to polyU database and results are analyzed in terms of receiver operating characteristic in this section. Section 5 summartise the chapter.

Keywords: radon transform, histogram of radon transform, fingerprint, and palm print

1. INTRODUCTION

The verification systems used in earlier days were knowledge based and token based. The forgotten or hacked password and lost key are the major drawbacks of these systems. Nowadays machine learning and artificial intelligence based processing of information is used in automation and in many sophisticated applications. Although many agencies have advanced the latest security systems, newest terrorist attacks exposed extreme weaknesses of sophisticated protection structures.

The verification of individual by the inherent characteristic of physical traits or behavioral traits classifies the individual from the inhabitants. This trait is also referred as modality. The most commonly used biometrics modalities are face, fingerprint, hand vein infrared thermogram, palm print, iris, gait, keystroke, ear, signature, voice, retinal scan etc.

1.1. Face

Face acknowledgment is a nonintrusive method, and facial pictures are presumably the most widely recognized biometric trait utilized. The location and the shape of the facial attributes are used for face recognition. Now-a-days lots of performance verification systems for face recognition are available [1].

1.2. Fingerprint

Fingerprint is a unique mark that standouts amongst the most respected biometric systems used to verify and recognize an individual. The ridges and valleys present on the exterior of the fingertip forms a particular pattern called as fingerprint (see Figure 1). The matching accuracy using fingerprints has been shown to be very high [2]. One issue with the present fingerprint detection frameworks is that they require a great deal of computational resources, especially when working in the conspicuous verification mode. Finally, fingerprints of few people might be inadmissible for customized recognizing verification because of inherited variables, developing, environmental, or word related reasons (e.g., manual masters may have innumerable cuts and wounds on their fingerprints that keep advancing).

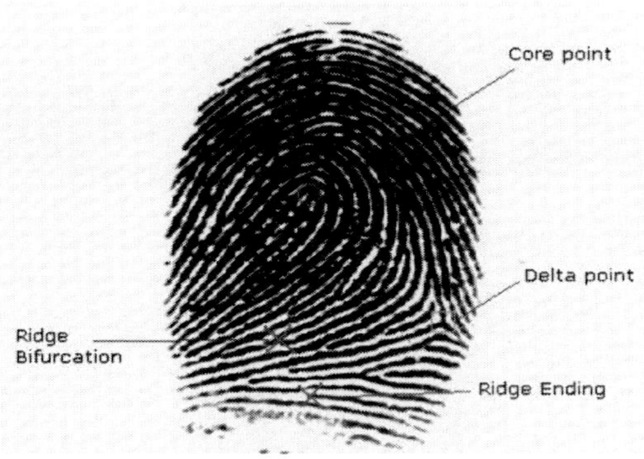

Figure 1. Singular points – A unique identification of fingerprint.

1.3. Hand Vein Infrared Thermogram

The infrared camera can be used to capture the pattern of radiated heat from human body. This pattern is a unique for individuals. Unlike the conventional methods, it is an unobtrusive technique.

1.4. Palmprint

Like the fingerprints, the palmprint of human beings also contains ridges and valleys. The palmprint gives bigger surface region for feature extraction as compared to the fingerprints. So the palmprints are more distinctive than the fingerprints. Moreover the palmprints have additional features like principle lines and wrinkles which can be used for identification and verification if an individual [3].

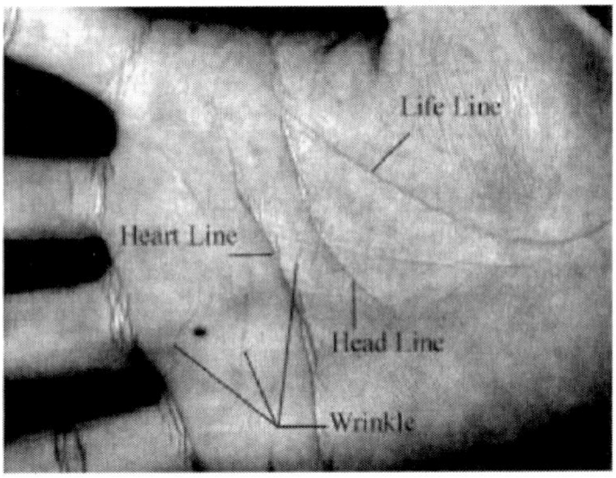

Figure 2. Unique identification of palmprint for each individual.

1.5. Iris

The iris is the annular area of the eye limited by the pupil and the white of the eye on either side. The complex iris surface conveys particular data helpful for individual recognition. Each iris is distinctive and, like fingerprints, even the irises of identical twins are different.

1.6. Gait

Gait is a particular way of walking. It should be particular, however is adequately prejudicial to permit confirmation in some low-security applications. Gait is a social biometric and may not remain invariant, especially over a broad stretch of time, on account of changes in body weight, noteworthy injuries including joints or cerebrum, or as a result of intoxication.

1.7. Ear

The pinna structure are the characteristic features of individuals's ear. The distance of salient points on these pinnas from a particular landmark is used for person recognition.

The present scenario for biometric identification is to use bimodality for better classification accuracy by strengthening the security. Major used modalities are fingerprint, iris, face, palm and voice. It has been adopted at large scale in major organization. The selection of modality plays an important role in the system. The literatures suggest that the enrollment process and computational complexity are the major constraints in the identification process. Quality of modality is an important aspect as large degree of variation in the modality makes classification process challenging. For face recognition, large variation in facial expression, illumination variation during modality acquisition, pose variation, sensor noise, blurring effect are the major constraints. Similarly, iris based identification system's performance is affected due to motion blur, spectacles, dilation etc. In addition, the outdoor and uncontrolled environments also affect the performance of the system. Following Figure 3 presents the biometric share in the market [4].

All modality imposes certain constraints and all modality cannot satisfy the requirements in general for all organization and/or applications. Various factors like environment under which it will be used, need of application and its requirements along with accuracy and ease of utilization have to be considered while selecting the modality. As suggested, few applications need utilization of multiple modalities to ensure the better accuracy. Therefore, selection of right biometric print is massively important to get the full benefits of the entire system.

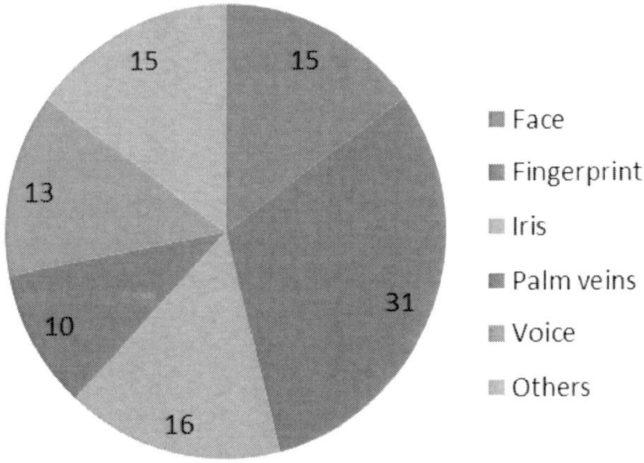

Figure 3. Major modalities used in market for biometric recognition (adapted from [4]).

Majority of the biometric systems are ranked using the two qualitative parameter entitled false acceptance ration (FAR) and false rejection ration (FRR). FAR represents number of times the query modality matches inaccurately with stored database. FRR gives rejection rate when the query modality is matched with one of the samples from database. The trade-off between the FAR and FRR represents the region of operating curve (ROC).

Another important parameter is equal error rate (EER). It is the average error obtained when FAR and FRR are equal. Both accuracy as well as EERmeasures the performance of biometric recognition system. The EER defines the ratio of the misclassified number of individuals and the total number ofmodalities' samples in the test set.

Figure 4 presents that Face based recognition is highly sensitive to the expression and pose and hence the FAR and FRR is quite high in comparison with other available modalities. Similarly voice traits are also vulnerable and in practice it is not preferable. Iris offers most secure recognition system with complexity of acquisition process. Lastly, Fingerprint and palmprint based biometric systems are more suitable due to ease of capturing the biometric process. Therefore this chapter processes the palmprint and finger print based biometric recognition system. Following Table 1 [4] makes comparative analysis of modality based on the few parameters.

Table 1. Comparison of modalities

Modality	Accuracy	Ease of Use	User acceptance
Face	Low	High	High
Fingerprint	High	Medium	Low
Iris	High	Medium	Medium
Palmprint	High	High	Medium
Voice	Medium	High	High

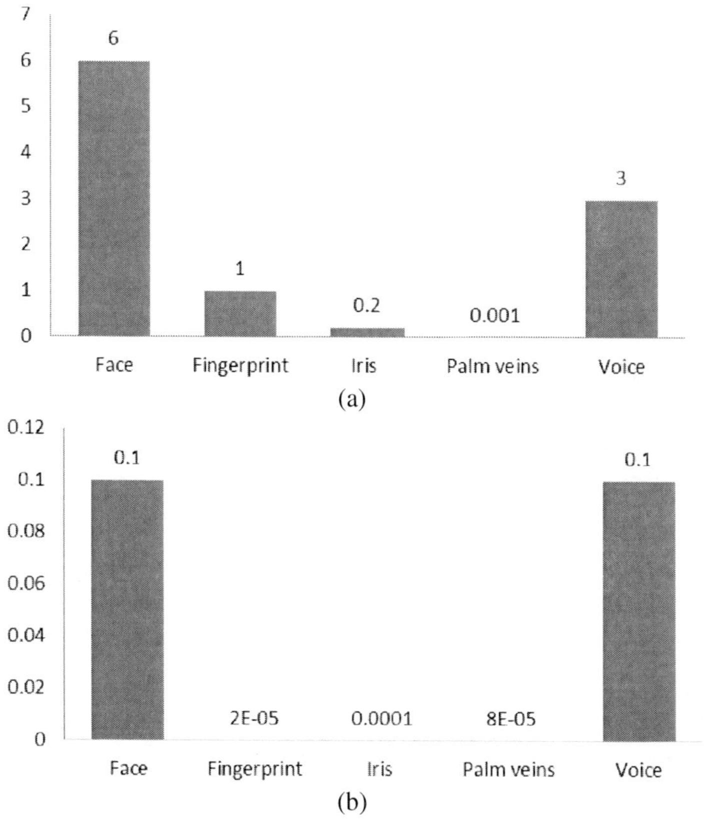

Figure 4. (a) False rejection ration and (b) False acceptance ration for modalities.

The basic block diagram of person identification system using unimodal biometric is shown in Figure 5. The system contains five components: (a) modality acquisition system using sensor or scanner, (b) preprocessing of the captured modality i.e., noise reduction, enhancement of image, region of interest (ROI) extraction etc. (c) extracting unique feature to be used for matching purpose (d) Matching process where extracted features are compared with stored features also referred as database.

Collection of large number of biometric impressions from the massive variety of users weaken unimodal biometric structures. Few limitations are enlisted below:

- A huge collection of biomtric impression weaken the unimodal biometric system and limitation imposes in the system like: artificial fingerprints can be created using the material and it had the option to impersonate more than one of every five fingerprints in a biometric system. This phony hand is perceived by methods for the HandKey II machine.
- By method of putting people finger in impression material, molds can be established of idle fingerprints through photographic scratching methodologies like the ones used in creation of PCB (sticky fingers) or play-doh and gelatin might be utilized to cast a phony finger.
- In worst case state of affairs lifeless hands may be supplied to the biometric recognition gadget [5].

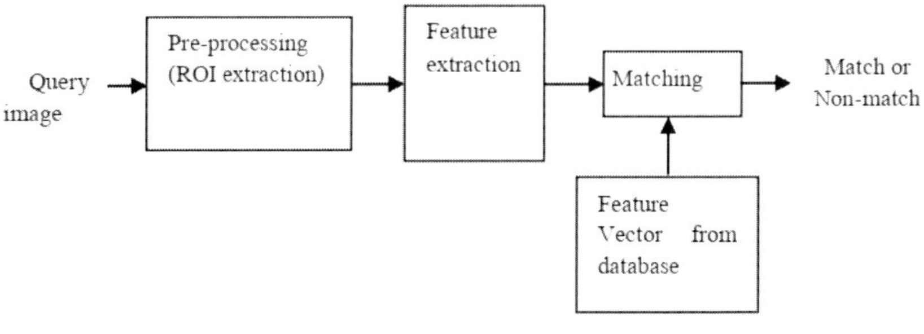

Figure 5. Unimodal biometric identification system.

The several biometric framework joins at least two distinctive biometrics of an individual (for ex. fingerprint and palmprint, face and unique finger impression, iris and unique finger impression, and so on.)

detected by utilizing different sensors. The use of multiple traits is also referred as multi model biometric system.

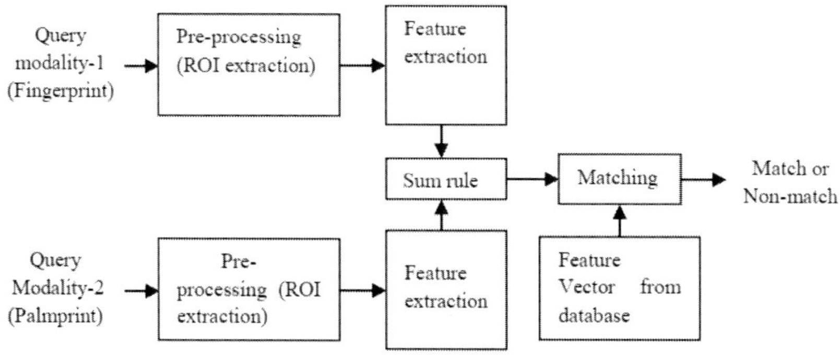

Figure 6. Bimodal person recognition system.

Use of multiple traits requires integration at different components which includes integration of traits at sensor level, integration of features, integration at matching components or integration of score at decision components. Using the multiple traits the biometric system can be represented as shown in Figure 6 where a sum rule is used to integrate the features of two modalities. During enrollment phase, the fused features from two modalitieshave been stored in the database. In the identification process, the features of query images are extracted. Later integrated features of these modalities are used in matching component for identification purpose by comparing them in the database.

In: Nova Excerpts: Earth Sciences
Editor: Dominik Lucas
ISBN: 978-1-53619-038-0

Chapter 8

RADON EXPOSURE: HEALTH EFFECTS AND PROTECTION*

Esra Kavaz and Neslihan Ekinci†
Department of Physics, Ataturk University, Erzurum, Turkey

ABSTRACT

It is known that approximately 50% of the radiation, we were exposed to in daily life, was caused by Rn isotopes, especially from 222Rn. Radon gas formed as a result of the degradation of 238U in the rocks is spread to the soil by diffusion and then to the atmosphere or environment. With the accumulation of radon gas, the density of Rn can reach critical values in confined spaces or in poorly ventilated areas. When radon gas is inhaled, densely ionizing alpha particles emitted by deposited short-lived decay products of radon can interact with biological tissue in the lungs leading to DNA damage. Therefore, exposure to radon and radon products is a potential public health problem.In this chapter, we focus on indoor radon exposure which originate from some building materials. Almost any building material made from natural materials, containing concrete and wallboard, might give off some level of radon. Radon gas emitted by soil

---

* The full version of this chapter is published in Radon*: Detection, Exposure and Control*, edited by Ramesh Chandra Tiwari, published by Nova Science Publishers, Inc., 2020.

† Corresponding Author's Email: nekinci@atauni.edu.tr.

or rock can enter buildings through cracks in floors or walls; construction joints; or gaps in foundations around pipes, wires, or pumps. Moreover, radon may be released at high concentrations from the water source to the air and can be inhaled as it moves from water to air. In addition, we will examine the radon risk levels in some occupational groups (like; mine, uranium processing factories, etc..). Furthermore, we present the general knowledge about the selection of radon measuring devices and the development of procedures for the reliable measurement of radon in both air and water in this section. Lastly, we discuss the strategies of countries about from radon prevention or to achieve an overall risk reduction.

Keywords: radon, health, exposure, protection

INTRODUCTION

Human beings and other living creatures have been beamed up by cosmic rays from the universe and radiation emitted from natural radioactive substances in the earth for millions of years and they have been constantly living together with radiation since the existence of all living things.

A small amount of radioactive substances in air, water, all vegetable and animal origin foods are taken into our body through respiration and digestion and these are accumulated in various organs over time. In addition, considering that influenced by cosmic rays and natural radioactive substances in the earth, the human body is naturally exposed to both internal and external radiation irradiations. The living creatures take the amount that is equivalent to two thirds of the radiation they receive from the natural sources, the air they breathe, the food they eat and the water they drink.

Nowadays, people spend much of their time mostly in indoor environments. Among the most preferred indoor areas where people spend most of their time are the houses and workplaces as fundamental niche. People intrinsically wish the environment they are in and the air they respire to be clean and healthy. However, there are plenty of sources polluting the indoor environment and the air in them. Radon is the most critical and the most dangerous source in indoor areas in terms of its effect on human health.

The most effective and common way to expose the human body to radon gas is by inhalation.

Since the radon concentration in the closed spaces is highly dependent on the geological conditions of the region, the geological structures of the regions where high concentrations are observed should be examined in terms of radon concentration sources. In addition to geological conditions, building materials, soil or water structure surrounding the area are also effective on indoor radon concentration. Depending on temperature and pressure difference, radon enters the buildings The floors and walls of dwellings contain many small cracks and gaps formed during construction. Radon from the ground is drawn into the building through these cracks and gaps because the atmospheric pressure inside the building is usually slightly lower than the pressure in the underlying soil. This small pressure difference is caused by the stack (or chimney) effect of heat in the building and by the effects of wind. Radon can also be found in the water we use and can be released into the air in your home when water is used for showers and other home uses.

In order to remove the radon from living areas, some protective measures should be taken. First, it is necessary to raise awareness of people about radon gas. Because radon gas makes the biggest contribution to the equivalent dose that people get from natural radiation sources. It should also be remembered that radon gas is one of the second biggest causes of lung cancer after smoking.

The reactivity of radon is weak. Therefore, it does not chemically bind to tissues when inhaled.

Also the solubility in the tissues is very low. However, the decay products of the radon attach to dust and other particles to form radioactive aerosols. Therefore, they could be taken through respiration by being transported. Decay continues until the decay products become stable; radiation releases at every stage of the decay process. The radiation dose increases in the bronchi epithelium as a result of the decay on the bronchial tube.

Since some of the decay products are alpha-emitters, the biological effects of alpha radioactivity will be gained.

When radon is released, the ionizing alpha particles absorbed by the short-lived decay products of the radon can lead to a breakdown of the DNA by interacting with the biological tissue in the lungs (Zeeb and Shannoun 2009). Given that even a single alpha particle can cause large genetic damage to cells, it can be seen that DNA degradation associated with radon is likely to occur at any level of exposure. Therefore, it is not possible to define a concentration level at which the radon will not cause potential lung cancer and its lower levels as a limit value. Studies on radiation-induced lung cancer cases in recent years are mostly related to the association of lung air and radon concentration in indoor air, and these studies have proven the accuracy of this suspicion. The spread of the knowledge that radon has been everywhere since the mid-1980s has brought a new perspective to the subject of radiation. Since then, research has been accelerated to measure radon density especially in indoor environments.

In order to control the radon density in indoor environments, countries and international authorities have imposed several restrictions on the highest levels of radon levels to be allowed in these environments. In buildings where these limits are exceeded, it is recommended or obliged to take measures to reduce the concentration of radon in the building.

In the number of 65 report of ICRP on the protection of radon and radon products in the home and workplaces, the effect of inhalation of radon and radon products on human health was investigated and the intervention limits for the annual effective dose to be taken from radon were determined to be 3-10mSv (IRCP 1993).

The limits applicable to our country and reported by TAEK are; 400Bq/m^3 for homes and 1000Bq/m^3 for workplaces (Official Gazette 2004). The European Union is based on 400Bq/m^3 in old buildings and 200Bq/m^3 in new buildings (TAEK 2008).

ICRP, however, reported that the levels of intervention in dwellings were 200-600Bq/m^3 and these limits for workplaces were 500 - 1500Bq/m^3, but within the framework of the latest scientific evaluations, WHO decided to reduce the number of health problems caused by exposure to radon gas. It recommends applying the reference level Bq/m^3 (Zeeb and Shannoun 2009). In cases where this level cannot be applied due to the special circumstances of the countries, it is emphasized that an in-house reference level which does not exceed the limit of at least 300Bq/m^3 should be determined.

The UK National Radiological Protection Council considers the locations where the radon measurements in NRPB homes exceed 200Bq/m^3 (Radon affected area) as 1% of the radon measurements (Denman et al. 1998).

General Properties

General characteristics of radon are briefly summarized in Table 1. Radon is a colourless, odourless, and tasteless gas and therefore is not detectable by human senses. The atomic radius of radon is 1.34 angstroms and it is the heaviest known gas. It is nine times denser than air. Radon is a single atom gas unlike oxygen, O^2, which is comprised of two atoms. Radon easily penetrates many common materials like paper, leather, low-density plastic (like plastic bags, etc.) most paints, and building materials like gypsum board (sheetrock), concrete block, mortar, sheathing paper (tar paper), wood panelling, and most insulations.

Radon is fairly soluble in organic solvents and water. The reaction with other compounds is comparatively limited. Moreover, radon is not completely inert and constitutes stable molecules with highly electronegative materials. Radon can be considered as a noble gas that occurs in several isotopic forms.

$Radon^{222}$ and $Radon^{220}$ are found in major concentrations in the human environment. $Radon^{222}$ is a member of the radioactive decay chain of uranium238. Radon220 is formed in the decay chain of $thorium^{232}$. $Radon^{222}$ decays in a sequence of radionuclides called radon decay products, radon daughters, or radon progeny. $Radon^{222}$ gas is mostly formed in the environment.

$Radon^{222}$ is released into the atmosphere from decay products of heavy metals such as polonium, lead, and bismuth.It can rapidly attach to other airborne materials such as dust and other materials facilitating inhalation.

Table 1. Some properties of the Radon gas

Property	**For Radon**
Sym.	Rn
Atomic Number	86
AtomicWeight	222
Melting Point	-71.15 °C
Boiling Point	-61.85 °C
Density	9.73g/L
Electron Configuration	$[Xe]4f^{14}5d^{10}6s^{2}6p^{6}$

Table 2. ^{222}Rn chain decay

Radon-222 →α γ	Polonium-218 →α γ	Lead-214 →β γ	Bismuth-214 →β γ	Polonium-214 →α γ
3.8 days	3 min	27 min	20 min	0.2 ms
Lead-210 →β γ	Bismuth-210 →β γ	Polonium-210 →α γ	Lead-206	Lead-210 →β γ
22.3 years	5 days	138 days	stable	22.3 years

Radon Concentration Limits

Indoor radon measurements were first made in Sweden in 1956. Although radon gas is present in very high concentrations in some environments, but its importance is ignored and there is not much emphasis. This is considered to be a special exception to the measuring zone. However, after 20 years, large-scale systematic studies have been started in different countries all over the world. In order to control the radon gas concentration in indoor environments, limit values have been determined by both countries and international organizations. If the limit values are exceeded, it is recommended to take measures to reduce the radon concentration.

According to the International Atomic Energy Agency Basic Safety Standards (IAEA-BSS), the recommended levels for radon are 200-600Bq/m^3. The concentration of radon that is authorized in Turkey is 400Bq/m^3. Table 4 describes the permissible radon concentrations for indoor environments adopted by various countries and international organizations in the report released by UNESCEAR in 2000 (UNSCEAR 2000).

The Effect of Radon on Human Health

The most effective and known way to expose the human body to radon gas is by inhalation. When inhaled radon gas, ionizing alpha particles emitted by radon's short-lived degradation products may interfere with biological tissue in the lungs, causing DNA degradation.

Considering that even a single alpha particle causes a major genetic damage to the cells, it can be seen that even at least the exposure to radon may damage in DNA. Therefore, it is not possible to refer to a concentration level and its lower levels as a limit value in which radon does not cause potential lung cancer. Most people are exposed to low to moderate levels of radon, and among them, the incidence of lung cancer is greater than those exposed to high concentrations of radon. Therefore, it is not possible to determine a low concentration limit value where radon does not cause potential lung cancer. Often people are exposed to low or moderate radon levels. These people are more likely to develop lung cancer than those exposed to high radon concentrations.

Research has shown that radon gas - induced lung cancer cases are quite high in mine workers. In this paper, (Zhou et al. 2019) presented the study of radon concentration in a typical underground uranium mine in China. There are numerous studies the concerned with analyses of radon sources and factors affecting radon exhalation in underground uranium mines (Mudd 2008; Hu and Li 2012; Sahu et al. 2013, 2014, 2015,2016; Zhao et al. 2016; Hong et al. 2015, Jackson and Schwendiman 1980; Chen et al. 1999; Zhou et al. 2005; Somlai et al. 2006; Marusiakova et al. 2011; Xie et al. 2012; Zhang et al. 2013; Bekteshi et al. 2017).

Studies investigating the relationship between radon concentration and lung cancer in the indoor environments in Europe, North America and Asia have revealed radon as the cause of the majority of lung cancer cases. The proportion of cases of radon-induced lung cancer varies between 3% and 14%, depending on the average radon concentration and measurement methods available in the country. In some countries, the relationship between indoor radon concentration and lung cancer cases was examined and Table 3 below emerged (WHO 2009).

Table 3. Radon gas limit values (Bq/m^3) accepted by some countries and organizations

The USA	**150**	**India**	**150**	**Norway**	**200**
Germany	250	Britain	200	Russia	200
Australia	200	Ireland	200	Turkey	400
Chine	200	Swedish	200	EU*	400
Denmark	400	Canada	800	ICRP**	400
France	400	Luxembourg	250	WHO***	100

*European Union (EU).

**International Commission on Radiological Protection (ICRP).

***World HealthOrganization (WHO).

Annual Effective Dose Value

The equivalent annual effective dose (A.E.D.V) is defined as the radiation dose that people will receive for one year by being exposed to radiation from different sources of radiation, either internally or by exposure to radiation from different radionuclides in the air they eat and drink or breathe. The equivalent dosage unit is Sievert. In calculating the annual effective dose equivalent, the radiation dose conversion factors emitted per unit activity for each radionuclide are determined using the radiation type and energy emitted by each radionuclide. The annual activity dose equivalent is found by multiplying the determined concentration of activity per unit volume or unit weight by dose conversion factors.

A.E.D.V=Df x C

where Df is the conversion factor, and C is the concentration of activity detected per unit volume or unit weight.

Measurement Method

Radon measurement techniques can be analysed in two groups as active and passive (Urban et al. 1981). Active and passive radon measuring instruments and their general characteristics are presented in Table 4.

Table 4. The Radon measurement devices and characteristics

Detector Type	Method	Uncertainty (%)	Sampling Period	Cost
Alpha-track detector	Passive	10-25	1-12 Months	Low
Activated Charcoal Detector	Passive	10-30	2-7 days	Low
Electret Ion Chamber	Passive	8-15	5 days -1 year	Medium
Electronic Integrating Device	Active	~25	2 days-years	Medium
Continuous Radon Monitor	Active	~10	1 hour – years	High

Active Measurement Method

In the active measurement technique, instantaneous radon measurements are made and this method is the shortest way to determine the radon concentration in a closed environment.

Air sample is taken from the place where the radon concentration is determined and the sample concentration is calculated in a radiation counter. Ion chambers, scintillation counter, electrostatic collectors and filters are examples of instruments used in active radon measurements. The most important feature of active measuring radon detectors is that the amount of radon in air, soil, water and building materials can be measured with the necessary equipment support (soil-prop, water-kit). However, this technique, which is based on sampling and is often used for short-term measurements, is not a highly preferred method for determining atmospheric radon concentrations within the building. Because the radon concentration which can be influenced by the external parameters such as humidity, pressure and air temperature can show significant changes in short time periods.

Passive Measurement Method

In the passive measurement technique, nuclear trace detectors which are left for a long time in the environment where the radon concentration will be measured are utilized. In order to determine an average value in radon measurements, it is generally appropriate to use passive measurement technique. Such long-term measurements can be performed daily, monthly, seasonal or yearly. Nuclear trace detectors used in the passive measurement technique are formed from films made of plastic materials such as cellulose acetate, polycarbonate or allyl diglycol carbonate.

The measurement is based on the fact that the invisible traces left by the alpha particles hitting plastic plates are clarified by chemical etching by a microscope. It is reported that passive measurement technique can be used to determine radon gas concentration in indoor environments, caves, mines and tunnels. The passive measurement method generally includes thermoluminescence detectors (such as $CaSO4$: Dy or LiF), or solid state nuclear trace detectors; cellulose nitrate (LR-115) or allyl diglycol carbonate (CR-39). Radon dosimeters using passive alpha detectors can be created in 3 different ways:

1. Devices with an air gas flow where radon daughters are collected on a filter in front of the detector
2. Passive devices without gas flow in which α particles from radon/daughters from a given air volume or after plate out from the chamber surface are detected
3. Passive devices using a bare detector which registers radon and daughter product in the ambient air

Radon gas measuring devices and their characteristics are given in Table 4. The radon measurement devices given in this table are the standard devices used among World Health Organization (WHO) member countries (WHO 2009).

In: Nova Excerpts: Earth Sciences
Editor: Dominik Lucas
ISBN: 978-1-53619-038-0

Chapter 9

ESTIMATION OF RADON AND ENVIRONMENTAL RADIOACTIVITY*

Pankaj Bala[1,†], Vinod Kumar[1] and Rohit Mehra[2]
[1]Department of Applied Sciences, Humanities and Management, Beant College of Engineering and Technology, Gurdaspur, Punjab, India
[2]Department of Physics, Dr. B.R. Ambedkar National Institute of Technology, Jalandhar, Punjab, India

ABSTRACT

Radioactivity is a part of our physical environment, which mainly comprises of air, soil and water. The health of the general public has been adversely effected by the exposure of living beings to the natural radiation particularly from the radon and natural radionuclides. The naturally occurring radionuclides present in soil include Uranium, Radium, Thorium and Potassium. Since these radionuclides are not uniformly distributed, the knowledge of their distribution in soils and rock play an important role in radiation protection and measurement. The radon problem presents a

---

* The full version of this chapter is published in *Radon:* Detection, *Exposure and Control*, edited by Ramesh Chandra Tiwari, published by Nova Science Publishers, Inc., 2020.

† Corresponding Author's Email: pankajbala.nitj@gmail.com.

global concern of radiation hygiene to the world population. The measurement of radon is of great interest because of its alpha emitting nature. Recent studies of many researchers have reported comparatively high value of Radium, Thorium and Potassium in some locations. Along with that in order to check possible changes in radioactivity level with time due to human interference to the atmosphere, there is a need for the measurement of radon and natural radionuclides from radiation protection and health point of view.Various international agencies like UNSCEAR, ICRP, WHO, IAEA, US EPA, etc. are regularly monitoring the exposure of the public, its health effects and the radiation levels thereby are giving recommendation for protection from radiations.

Keywords: radioactivity, natural radionuclides, radon, health hazards

INTRODUCTION

Environment is the combination of those physical circumstances in which we live and is essential to our health. We need to improve our day-to-day life; however, every action of man for this improvement by over - exploitation of natural resources leads to environmental pollution. The purpose of environmental analysis is to determine the background natural concentrations of the identified chemical constituents in the environment. Radioactivity is the term used to describe disintegration of atoms. The atom can be characterized by the number of protons in the nucleus. The Radioactive decay is the process by which an unstable atomic nucleus loses energy by emitting radiation such as alpha particle, beta particle and gamma particle. A material containing such unstable nuclei is known as radioactive. Terrestrial radioactivity and the associated external exposure due to the gamma radiation depend primarily on the geological and geographical conditions and appear at different levels in the soils of each region (UNSCEAR 2000). The largest contribution to the radiation field is of natural origin: it is due to the cosmic rays, the natural radionuclides in soil and other building materials, radioactivity of the ground and the radioactive decay products of radon in the air. Naturally occurring building materials reflects the geological variations of their sites of origin, so their radiometric characterization provides a useful technique of acquiring better understanding of the local environment (Trevisi et al., 2005). Artificial radioactivity emitted from nuclear power plants, industrial plants and research facilities has smaller contribution to the overall radiation. These emissions are very small in normal operation, although large amounts of radioactivity can be released to the environment through accidents. All building materials and soil contains various amount of natural radioactive nuclides. Materials derived from rock and soil contains mainly natural radionuclides i.e., uranium (^{238}U), thorium (^{232}Th) and potassium (^{40}K). In the uranium series, the decay chain segment starting from radium (^{226}Ra) is radiologically the most important and, therefore, reference is often made to radium instead of uranium. The world wide average concentrations of radium, thorium, and potassium in the earth's crust are 33 $Bqkg^{-1}$, 45 $Bqkg^{-1}$ and 420 $Bqkg^{-1}$, respectively (UNSCEAR,2000). The concentrations of the natural radionuclides (^{238}U, ^{232}Th,their daughter products and ^{40}K) present in the soils and rocks depend upon the local geological and geographical conditions of each region in the world (Quindos et al., 1994, UNSCEAR 2000). Because of this reason, there are few high natural background areas in India and in world e.g., Yangjiang in China, Ramsar in Iran, Orissa coast and Kerala coast of India, etc. (Sunta et al., 1982 & 1993; Mishra et al., 1993; Sohrabi et al., 1993; Zhu et al., 1993; Ramola et al., 2010). Currently, a worldwide effort is underway to measure the activity concentration in building materials (UNSCEAR, 1988; Tufail et al.; 1992). A range of radioanalytical methods can be used for determination of activity concentrations in naturally occurring materials. Some radioactivity studies have been previously carried out in soil and building materials in some parts of the world (Beaza et al., 1992; Noordin, 1999; Selvasekarapandian et al., 2000; Kannan et al., 2002; Patra et al., 2006; Mehra et al., 2007; Senthilkumar et al., 2010; Bala et al., 2014). Therefore, it is necessary to measure the natural radioactivity in soil throughout the world, which led to worldwide national surveys in the last three decades (McAulay and Moran, 1988).

Radon is naturally occurring radioactive gas and comes from breakdown of Uranium (U^{238}) following by the alpha decay of radium isotope (Ra^{226}). It directly decays into polonium (Po) having energy of approx 5.5 MeV emitting alpha particles. In view of the fact that radon, thoron and their progeny concentrations contribute the most to the natural radiation dose to general populations, large scale and long-term measurement of radon, thoron and their progeny concentrations has been receiving considerable attention (Mayya et al., 1998). Even though measurement of radon, thoron and their progeny concentrations was done over the past 50 years in many countries, with the improvement of experimental apparatus and technical formulation, the same is going on till today. The measurement of radon, thoron and their progeny concentrations also leads to the knowledge of the presence of radioactive elements, which are the sources of these elements. Since Uranium-238 is the parent nuclei of Radon and Thorium-232 that of Thoron, hence with the concentrations of these gases in air, one can predict the presence of high or low concentrations of the source radioactive elements. Measurement of indoor radon is of importance because the radiation dose to human population due to inhalation of radon and its daughters contributes more than 50% of the total dose from natural sources (UNSCEAR 2000). During the recent years, several reports in literature have explained the ever-increasing interest in monitoring the radon in the dwellings all over the world (Abu-Jarad and Fremlin 1981; Keller and Folkerts 1984; Nazaroff and Doyle 1985; Ramachandran et al. 1986; Ramu et al. 1990; Virk et al. 1999, Bala and Mehra 2015). The results of the studies show that some countries (Sweden, Norway, Hungary and some parts of USA) have high radon concentrations in many of their dwellings and even in certain cases, doses from this source for some people living in these areas may exceed those received by occupational workers. In India, Ramu et al. (1990) had initiated indoor radon measurements in some dwellings of 15 towns which were specially selected on the basis of high uranium content in the soil. Radon monitoring has been done in the dwelling of Haryana (Kant and Chakravarti 2004). Studies of radon and thoron levels in Mysore City have also been done (Chandrashekara and Paramesh, 2008). In Mizoram, measurement of radon, thoron and their progeny concentrations were done only in 17 dwellings (Ramachandran et al., 2003).

There were number of expert bodies formulated in order to regulate the exposure to radiation. In 1928, an independent non-governmental body of experts in the field, the International X-ray and Radium Protection Committee was established. It later was renamed the International Commission on Radiological Protection (ICRP). Its purpose is to establish basic principles for, and issue recommendations on, radiation protection. These principles and recommendations form the basis for national regulations governing the exposure of radiation workers and members of the public. They also have been incorporated by the International Atomic Energy Agency (IAEA) into its Basic Safety Standards for Radiation Protection published jointly with the World Health Organization (WHO), International Labour Organization (ILO), and the OECD Nuclear Energy Agency (NEA). These standards are used worldwide to ensure safety and radiation protection of radiation workers and the general public. An intergovernmental body was formed in 1955 by the General Assembly of the United Nations as the United Nation Scientific Committee on the Effects of Atomic Radiation (UNSCEAR). UNSCEAR is directed to assemble study and disseminate information on observed levels of ionizing radiation and radioactivity (natural and man-made) in the environment, and on the effects of such radiation on man and the environment. Along with that in order to check possible changes in radioactivity level with time due to human interference to the atmosphere, there is a need for the measurement of radon and natural radionuclides from radiation protection and health point of view.

In: Nova Excerpts: Earth Sciences
Editor: Dominik Lucas
ISBN: 978-1-53619-038-0

Chapter 10

DYNAMICS OF ^{222}RN AND ^{220}RN DECAY PRODUCTS AND THEIR DEPOSITION BASED MEASUREMENT TECHNIQUE*

***Rosaline Mishra*†**
Radiological Physics and Advisory Division,
Bhabha Atomic Research Centre, Mumbai, India

ABSTRACT

The importance of measuring the decay products of ^{222}Rn (Radon) and ^{220}Rn (Thoron) is that, these are the major inhalation dose givers. More precisely, out of 1.29 mSv of annual effective dose due to inhalation of ^{222}Rn, ^{220}Rn and their decay products, ~95% is attributed solely due to inhalation of the decay products. These decay products, essentially the radioisotopes of Polonium, Bismuth and Lead are solid particulates, which deposit in different parts of the respiratory tract, subsequently emitting high energy alpha particles upon their radioactive decay. Development of passive deposition based direct progeny sensors known as Direct Radon and Thoron progeny sensors, have provided an easy-to-use technique for

* The full version of this chapter is published in Radon*: Detection, Exposure and Control*, edited by Ramesh Chandra Tiwari, published by Nova Science Publishers, Inc., 2020.

† Corresponding Author's Email: rosaline.mishra@gmail.com

time integrated measurements of the decay products only. These dosimeters are apt for large scale population dosimetry to assign inhalation doses to the public. It is being also used in the occupational environments of Uranium mines and Thorium plants as personnel dosimeters. The chapter gives an insight into the dynamics of the progeny of ^{222}Rn and ^{220}Rn and its direct measurement technique.

Keywords: Direct Progeny sensors, progeny, inhalation dose, DTPS, DRPS, Radon, Thoron, unattached, aerosol

INTRODUCTION

The studies on the health effects of acute radiation exposures among the atomic Bomb survivors of Hiroshima and Nagasaki and the 11 mine combined study of uranium miners exposed to high levels of radon are among the foremost studies providing quantitative estimates of the radiation risk coefficients. The former provides risk estimates at relatively high doses of gamma radiations and neutrons and the latter studies principally establish the lung cancer risks due to alpha particle radiations from the deposited decay products of radon. Although not directly applicable to low doses, these have formed the basis of ICRP dose limits for occupational workers and general public, through a low dose extrapolation technique based on LNT hypothesis. To understand the nature of radiation risks directly at low doses, several studies were mooted. Notable among these are the pooled studies using the data of radiation workers from several countries and residential radon epidemiology. Until recent times, these studies did not yield statistically significant risks and results were generally equivocal. However, in the last two years, major publications have been brought out reporting statistically significant evidences of for low dose risks. The study by Cardis et al. (2007) conducted by pooling a database of workers from 15 countries showed that radiation risks could be comparable (if not higher) to that extrapolated from A-Bomb survivor data. Similarly, in the case of radon, the European case control (Darby et al. 2005) and North American pooling studies (Krewski etal., 2005) have yielded evidence of lung cancer risks at the rate of 16% excess RR per 100 Bq/m^3 due to residential radon exposures. These risks are at levels comparable to that predicted by extrapolating miner's data.

Considering that these studies are essentially based on western populations, and the studies on the populations of High Back Radiation Areas in China and India have not shown evidences of higher risks, questions are being asked regarding the validity of these risks across populations. This is an important issue that calls for serious examination of radiation effects at low doses among various population groups. In addition, although these large population-based epidemiologic studies conducted over the past 20 years, in North America, Europe, and China were solely based on measurement of exposure due to indoor radon. The recent North American and European pooled residential radon epidemiologic studies, which is essentially based on radon-only measurements, reported increased risks of lung cancer (Darby et al., 2005; Darby et al. 2006; Krewski et al. 2005; Krewski et al. 2006). But then these pooled risk estimates were not based on the radon progeny exposure that actually deliver the radiologically significant dose to the lungs. The reason is the availability of a large variety of instruments to measure Radon gas, but the unavailability of a suitable passive technique for measurement of long term time integrated progeny measurement. The progeny of Radon and Thoron, which are the actual dose-givers are particulates unlike their parents ^{222}Rn and ^{220}Rn that are gaseous in nature. Their formation and properties are discussed in the subsequent sections.

Radon, Thoron and Their Progeny

In the three primordial radioactive chains ^{238}U, ^{232}Th and ^{219}Rn, three radon isotopes ^{222}Rn, ^{220}Rn and ^{219}Rn are created. Within the natural uranium isotopes, the relative abundance by weight is respectively 99.28% for ^{238}U and 0.71% for ^{235}U. Hence ^{238}U predominates over ^{235}U in the environment. So, ^{219}Rn (formed in the decay chain of ^{235}U), is generally ignored because of its short half-life (4 s) and the comparative scarcity of ^{235}U. Half life of ^{220}Rn (55 s) also restricts its distance of movement from the source before its decay. Though the chemistry of ^{222}Rn and ^{220}Rn are the same, the difference in their half-lives 3.82 days and 55 seconds respectively, solely governs the differences in their physical behavior and the effects.

Both ^{222}Rn and ^{220}Rn decay by alpha particle emission to form radon decay products/ progeny which can be divided into the short-lived (half-lives <30 min) and long-lived progeny species.

Progeny Formation and Dynamics

When ^{222}Rn decays emitting ^{218}Po and an alpha particle is emitted. The kinetic energy of the alpha particle is never quite equal to the disintegration energy Q because, since momentum is conserved, the nucleus recoils with a small amount of kinetic energy when alpha particle emerges (Porstendörfer 1984). The excess electrons remaining after the alpha emission in the daughter nuclide are stripped away during recoil. Usually more than 1 orbital electrons is lost making ^{218}Po positively charged at birth. (Stripping of orbital electrons during the recoil: Bohr Linhard Principle). So, 90% of the daughter nuclide atoms acquire +ve charge and 10% remain neutral. The diffusion coefficient of the charged species (0.07-0.08 cm^2/s) is less than that of neutral species (0.025-0.05 cm^2/s) because the charged species undergoes multiple interactions with the surrounding aerosol atoms (Chu and Hopke 1988; Hopke 1996; Porstendörfer and Reineking 1992),

Cluster Formation and Neutralisation

Freshly generated radionuclides react fast (<1s) with the trace gases and air vapours and become small particles called clusters or unattached fractions (0.5-5nm diameter). This cluster formation is accompanied by the neutralisation process, which is essentially described by three mechanisms (Goldstein and Hopke 1985; Chu and Hopke 1988).

Charge Transfer with Airborne Neutral Atoms

(^{218}Po+ being unstable, it may react chemically with the air oxygen to form polonium dioxide (PoO^{+2}) for which they estimated an ionisation potential of ~10 eV. In such circumstances, the polonium dioxide can become neutralised by removing electrons from trace gases having a substantially lower ionisation potential (such as NO, NO_2, NH_3 or SO_2) and/or water vapour.)

Electron-Scavenging by Means of the Airborne OH Radicals

Radiolysis of the water and/or alcohol molecules leads to formation of the hydroxyl (OH) radicals, which are good electron acceptors (with an electron affinity of 1.83 eV). The airborne free electrons, are originated from the gas ionisation due to the parent (^{222}Rn) decay. Accordingly, the formation of airborne OH radicals plays an important role both in the neutralisation process for the polonium ion).

Recombination with Airborne Small Negative Ions

(In this mechanism, the airborne ^{218}Po ions may recombine with small negative ions (including electrons). These last are produced as a result of the air ionisation from α-emission as well as the ^{222}Rn and/or ^{220}Rn daughter recoil process).

Attachment to Aerosols

The process by which the unattached Progeny of ^{222}Rn and ^{220}Rn undergo random motion like gas molecules stick to the aerosol particles is called the process of attachment. It is a first order process.(Dankelmann et al, 2001; Hopke 1996; Pagelkopf and Porstendörfer 2003; Papastefanou 2008),Given $\Delta Z(d_p) = Z(d_p)\Delta(d_p)$ the aerosol particle concentration (in units of cm^{-3}) in the size interval Δdp of the particle diameter dp and number concentration $Z(d_p)$, the aerosol-attachment rate, $\lambda_a(d_p)$ (h^{-1}), that expresses the adsorption velocity of the unattached decay products to aerosol particle in air is given by

$$\lambda_a(d_p) = \beta(d_p)\Delta Z(d_p) = \beta(d_p)Z(d_p)\Delta(d_p) \quad (1)$$

where $\beta(d_p)$ ($cm^3\ h^{-1}$), which is a function of particle aerosol size d_p, is known as the aerosol attachment coefficient. The theory of aerosol-attachment is based on random collisions that occur between the airborne-unattached ^{222}Rn and/or ^{220}Rn progeny, aerosol particles, and gas molecules. Hence, the aerosol-attachment coefficient is satisfactorily described by the following expression

$$\beta(d) = \frac{2\pi D_0 d}{\frac{8D_0}{dv_0} + \frac{d}{d + {l_0}/{2}}} \tag{2}$$

where v_0 is the mean thermal velocity of the airborne-unattached ^{222}Rn and ^{220}Rn progeny (1.72×10^4 cm s^{-1} at NTP). There are two limiting values to equation

(1) For small particles ($d_p << l_0$), the aerosol-attachment is controlled by pure kinetic theory of gases and is proportional to the aerosol surface: $\beta(d) = \frac{\pi v_0 d^2}{4}$

(2) For large particles ($d_p >> l_0$), the aerosol-attachment process is determined by pure diffusion laws: $\beta(d) = 2\pi D_0 d$

Usual assumptions taken in the above theories are that i) the size and mass of the airborne-unattached ^{222}Rn and ^{220}Rn daughters are negligibly small compared to those of an aerosol particle, ii) the electrostatic forces need not be taken into account, and iii) the airborne-unattached decay products concentration is low enough that the corresponding agglomeration can be neglected.

A single value of the aerosol-attachment coefficient is, therefore, valid only for monodisperse aerosols. In practice, real aerosol particles in home, buildings and workplaces are polydisperse. Attachment rate indoors without any additional sources was measured as 20-50 h^{-1} (Mishra et al.,2009), and the attachment coefficient indoor was measured as $5*10^{-3}$ cm^3/h.

Desorption from Aerosol Particle

When recoil energy is higher compared to the potential energy of a molecule adsorbed on the surface then desorption occurs. Recoil factor of 0.8 is taken for ^{214}Pb. Upon decay of aerosol-attached ^{222}Rn and/or ^{220}Rn progeny by means of β- emissions, the resulting atoms may also become unattached on recoil. Nevertheless, their recoil is much smaller than that of α-emissions and no experimental evidence has been found to support the aerosol detachment after β-decay. The β-recoil fraction is therefore taken to be zero.

Size Distribution

The attachment leads to the distribution of progeny sizes: Unattached fraction (0.05-5nm and attached fraction (10-1000nm). Once the attachment is over the progeny becomes a part of the indoor aerosol. Now the progeny concentration indoors can either be reduced by transportation to outdoor by ventilation or by deposition on indoor surfaces.

Deposition on Surfaces

Airborne ^{222}Rn and ^{220}Rn decay products, whether unattached or attached to aerosol particles, tend to deposit on any available surface in direct contact to indoor air, becoming no longer available for inhalation and causing lung doses. The term plate-out is often used for surface deposition process of airborne ^{222}Rn and/or ^{220}Rn progeny. Following plate-out, re-entrainment into the air owing to α-recoil mechanism is neglected in most cases. Assuming always well-mixed indoor air and that all the surface areas in the reference room are equally effective in collecting ^{222}Rn and/or ^{220}Rn decay products, the respective loss rates by plate-out effect of these nuclides in airborne-unattached and aerosol-attached state are, respectively, given by Porstendörfer 1984):

$$\lambda_d^u = v_d^u \frac{S}{V} \text{ and } \lambda_d^a = v_d^a \frac{S}{V} \tag{3}$$

where the superscripts u and a refer to airborne-unattached and aerosol-attached state of the considered daughter atom, respectively, S includes the area of the walls, floor, ceiling and furniture in the reference room available for deposition, V is the indoor air volume and v_d^u and v_d^a are the deposition velocity of airborne-unattached and aerosol-attached ^{222}Rn and/or ^{220}Rn progeny.

Deposition of aerosol particles is governed by the physical processes of sedimentation by gravity, impaction by inertial forces, interception and Brownian diffusion. The deposition mechanism of particles on surfaces from turbulent air flow can be divided in to 2 types:

Transport towards the surface, determined by eddy diffusivity and

Precipitation on surfaces, a function of particle diameter, the structure and roughness of the surface.

The steady state particle flux towards the surface can be described by

$$\phi(d) = \left[D_T(z) + D_p(d)\right]\frac{dZ(z,d)}{dz} + v_S(d)Z(z,d)$$

where,

Z(z, d) is the concentration of particles with diameter *d* at a height *z* above the surface.

Fraction of flux by turbulent diffusivity $D_T(z)dZ/dz$ is much greater than the fraction by Brownian diffusion $D_p(d)dZ/dz$ and sedimentation $v_S(d)Z(z,d)$, where D_T and D_P are the turbulent diffusion coefficient and the Brownian diffusion coefficient respectively, v_s is the settling velocity of the particle.

A distinguishing feature of aerosol particles as opposed to trace gases is their ability to adhere to surfaces upon contact mainly because of the short range London-van der waal-Hamaker forces. As a result, particles transported to the surface (by molecular diffusion and turbulence in fluid motion) to within a few nanometer distances, deposit on them. The deposition mechanism is of significance in almost all applications of aerosols: lung deposition, removal in outdoor and indoor environments, nuclear aerosol plate-out in transport and containment systems, radon progeny equilibrium, pollution control strategies, microelectronics applications and material synthesis. Thus, understanding the mechanics of particle deposition is of vital importance in aerosol research.

Deposition of airborne nuclides on surface is parameterized in terms of deposition velocity which is expressed as (v_d):

$$v_d = \frac{\phi(d)}{Z(z,d)} \tag{5}$$

where,

$\phi(d)$ is the number of particles with diameter *d* deposited per unit surface and time.

It is estimated as a ratio of time integrated flux of atoms on the absorber surface to the mean airborne concentration of progeny. Regarding the unattached ^{222}Rn /^{220}Rn daughter clusters with a size of about 0.5-4 nm, the deposition velocity is about 100 times higher than the value of the aerosol attached radionuclides (diameter range 100-1000 nm).

Deposition velocity mainly depend on the particle size, aerosol concentration, ventilation rate, turbulence etc.

Progeny Concentration Indoors and Jacobi Model

In contrast to ^{222}Rn and ^{220}Rn, which are not influenced by the air constitution and characteristics – the nature and behaviour of their airborne progeny is further complicated due mainly to the diverse interactions that they experiment with the surrounding environment

^{222}Rn and/or ^{220}Rn progeny being in airborne-unattached form may participate in the following processes:

- to attach to any neighbouring aerosol particles,
- to deposit on the available surfaces (walls, floor, ceiling, and furniture),
- to be exhaust throughout the outdoor air by means of ventilation, and
- to decay to the following daughter.

On the other hand, the aerosol attached ^{222}Rn and/or ^{220}Rn decay products may have the following alternatives:

- to detach from their host aerosol because to the recoil energy acquired along the decayof their precursor, especially in the case of α-emissions.
- to deposit on the available surfaces (walls, floor, ceiling, and furniture),

- to be exhaust throughout the outdoor air by means of ventilation, and
- to decay to the following daughter.

A theoretical approach, which is widely used to predict the indoor ^{222}Rn and ^{220}Rn progeny behaviour, is that developed by Jacobi (1972). In this approach, the time dependence of the airborne ^{222}Rn and ^{220}Rn daughter concentration within a reference room is given by the balance between the generation and removal processes.

The rate of progress of each process is proportional to the concentration of the precursor species for that process. Thus, all the essential information is contained in a set of proportionality rate constants. Assuming complete air mixing and equal ventilation, aerosol-attachment and deposition rates for all ^{222}Rn and ^{220}Rn decay products, the set of differential equations predicting their indoor concentration dynamics among their possible states are given below:

For radon daughters:

$$\begin{aligned}
C_1^f &= \frac{C_0\lambda_1}{\lambda_1+\lambda_v+\lambda_w^f+X} & C_1^c &= \frac{XC_1^f}{\lambda_1+\lambda_v+\lambda_w^c} \\
C_2^f &= \frac{C_1^f\lambda_2+\lambda_2 rC_1^c}{\lambda_2+\lambda_v+\lambda_w^f+X} & C_2^c &= \frac{XC_2^f+\lambda_2(1-r)C_1^c}{\lambda_2+\lambda_v+\lambda_w^c} \\
C_3^f &= \frac{C_2^f\lambda_3}{\lambda_3+\lambda_v+\lambda_w^f+X} & C_3^c &= \frac{C_2^c\lambda_3+XC_3^f}{\lambda_3+\lambda_v+\lambda_w^c}
\end{aligned} \tag{6}$$

$$\begin{aligned}
C_1^f &= finefraction^{218}Po \\
C_1^c &= coarsefraction^{218}Po \\
C_2^f &= inefraction^{214}Pb \\
C_2^c &= coarsefraction^{214}Pb \\
C_3^f &= inefraction^{214}Bi \\
C_2^c &= coarsefraction^{214}Bi
\end{aligned} \tag{7}$$

Different rate processes such as the radioactive decay (λ_1, λ_2, λ_3), attachment rate (X), wall-loss rates for fine (λ_w^f) and coarse (λ_w^c) fractions and ventilation rate (λ_v),recoil factor (r) are incorporated. The unattached fraction (p1,p2,p3) for individual radon progeny is given by:

$$p_1 = \frac{C_1^f}{C_1^f + C_1^c}$$

$$p_2 = \frac{C_2^f}{C_2^f + C_2^c}$$

$$p_3 = \frac{C_3^f}{C_3^f + C_3^c}$$

And for thoron daughters

$$C_2^f = \frac{C_0 \lambda_2}{\lambda_2 + \lambda_v + \lambda_w^f + X} \quad C_2^c = \frac{XC_2^f}{\lambda_2 + \lambda_v + \lambda_w^c}$$

$$C_3^f = \frac{C_2^f \lambda_3}{\lambda_3 + \lambda_v + \lambda_w^f + X} \quad C_3^c = \frac{C_2^c \lambda_3 + XC_3^f}{\lambda_3 + \lambda_v + \lambda_w^c}$$

$$\begin{aligned} C_2^f &= finefraction^{212}Pb \\ C_2^c &= coarsefraction^{212}Pb \\ C_3^f &= finefraction^{212}Bi \\ C_2^c &= coarsefraction^{212}Bi \end{aligned} \tag{8}$$

Different rate processes such as the radioactive decay (λ_2, λ_3), attachment rate (X), wall-loss rates for fine (λ_w^f) and coarse (λ_w^c) fractions and ventilation rate (λ_v) are used. The unattached fraction is given by:

$$p_2 = \frac{C_2^f}{C_2^f + C_2^c} \tag{9}$$

$$p_3 = \frac{C_3^f}{C_3^f + C_3^c}$$

Direct Progeny Measurement Techniques

Conventionally, the dosimetry is carried out using radon and thoron gas concentration measurements and doses have been assigned using assumed equilibrium factors for the progeny species. The gas based dose assignment is somewhat inadequate due to variations in equilibrium factors and possibly due to significant thoron. In fact, since the true exposures depend upon the intricate mechanisms of progeny deposition in the lung, therefore an integrated approach for the assessment of progeny is essential. In this context, the recently developed Deposition based progeny concentration measurement techniques (DTPS: Direct Thoron progeny sensors and DRPS: Direct Radon progeny sensors) appear to be best suited for radiological risk assessments both among occupational workers and general study populations. These deposition based progeny sensors can be arranged in different modes to measure various progeny related parameters. These extracted multiple parameters give a better perspective while assigning the inhalation doses.

Another point against the use of gas concentrations as an index of exposure, is due to the growing realization on the importance of thoron. It is now being increasingly recognized as an unjustly neglected component in environmental radioactivity studies. However, due to its short half-life, thoron gas is not uniform in the room and hence its measurement is only an indicator of its presence and not a quantitative measure of the progeny concentrations.

There have been important developments, towards developing techniques for assessing progeny concentrations directly, wherein the deposition property of the progeny species has been exploited to relate the deposited flux to the activity concentration of progeny in the environment. In this regard, Direct Thoron and Radon progeny sensors (DTPS and DRPS) (Mishra and Mayya 2008) were developed, which provides a solution to the long standing question of personal dosimetry of subjects exposed to radon and thoron decay products, analogous to TLDs for Gamma radiations.

In: Nova Excerpts: Earth Sciences
Editor: Dominik Lucas
ISBN: 978-1-53619-038-0

Chapter 11

RADON RISK AND REMEDIATION: A PSYCHOLOGICAL PERSPECTIVE*

David Hevey†*, PhD*
School of Psychology, Trinity College Dublin, Dublin, Ireland

ABSTRACT

Although radon exposure in the home increases the risk of lung cancer, this risk can be managed. However, evidence indicates that testing for radon and subsequent home remediation rates are generally low in many countries. The present chapter outlines some key insights from psychological science that might account for sub-optimal radon protection. Psychological aspects of how the health risks posed by radon are perceived and managed are outlined. There is need to consider radon risk perception in terms of the (a) the cognitive and emotional responses to radon, and (b) the social context in which the radon threat occurs. In addition, the nature of the threat itself is integral to the failure for people to act in response to a radon threat. Finally, the challenges arising from defensive processing of radon threat information are outlined.

* The full version of this chapter is published in *Radon: Detection, Exposure and Control*, edited by Ramesh Chandra Tiwari, published by Nova Science Publishers, Inc., 2020.

† Corresponding Author's Email: heveydt@tcd.ie.

Keywords: radon, health threat, risk perception, heuristics, defensive processing

INTRODUCTION

You and everyone who lives in your home may be at increased risk of developing lung cancer, one of the most distressing, painful and fatal cancers. However, you can easily test to see if you and your family are at risk, and if you are at risk then you take action to effectively reduce the risk.

What would you do? Would you get a test? If the test came back telling you that you are at risk, would you do anything to reduce the risk?

For many people, their responses to this scenario are very straightforward – they tell you that of course they would test and that if they were at risk then they would then act to reduce the risk. However, if we add in details regarding the nature of cause of the risk – radon - we see overwhelming evidence that the actual behaviours both in terms of testing and remediation action to reduce the risk are not routinely performed. The high degree of variation in testing and remediation rates documented in the literature can seem counter-intuitive: why would you not test to see if you are at risk for something as severe as lung cancer? Why would you not reduce the risk either for yourself or your family/loved ones?

This chapter examines radon testing and remediation from a psychological perspective. Lung cancer is the leading cause of cancer-related deaths worldwide. Following tobacco smoke, exposure to radon is the second leading cause of lung cancer (WHO, 2009). The carcinogen radon is a naturally occurring radioactive gas that can accumulate in enclosed spaces, including homes. A review of inadequate housing in the EU region reported that radon exposure is associated with more than 3,000 deaths per year in the three countries of France, Germany and Switzerland (Braubach et al. 2011). The level of radon can be easily measured, and homes with high levels can be remediated to effectively reduce the associated risk; mitigation strategies typically focus on sealing radon entry routes, on ventilation and on reversing the air pressure differences between the indoor occupied space and the soil underneath the building, employing different soil depressurization techniques (WHO, 2018). A combination of strategies may be required to ensure effective remediation.

Research evidence consistently reports low levels of radon testing and home remediation. Despite high levels of awareness of radon being documented among general populations (Larsson et al. 2009), many people underestimate the seriousness and/or long-term health effects of radon exposure (Hill et al. 2006). In addition, even when individuals are informed that their homes have high radon levels, remediation rates can be very low, e.g., 7% (Kennedy et al. 1991). Comprehensive multi-media information programmes to increase radon testing and remediation internationally are successful at increasing awareness of radon; however, research consistently finds low levels of radon testing and remediation following such programmes (e.g., Nissen et al. 2012). The lack of appropriate testing and remediation by individuals poses profound public health challenges. It is in this context that the present review examines the psychological aspects of how the health risks posed by radon are perceived and managed.

Managing environmental risks, such as radon, can be achieved by government legislation alone, individual action alone, or some combination of legislation and individual action. In many countries, radon testing and remediation are the responsibility of the individual: consequently, in such contexts the goal of public awareness communications is to help individuals take appropriate preventive action. The individual is responsible for (1) testing to determine to what extent radon is present, (2) deciding if the level poses a threat, (3) selecting an appropriate remediation strategy, (4) implementing the remediation strategy, and (5) retesting to ensure that remediation has been successful. To ensure that the individual has the requisite knowledge to make an informed decision regarding radon, the government's role is typically to provide information to individuals regarding the threat radon poses, its assessment and potential remediation strategies. In essence, the state communicates rather than regulates: action regarding testing and remediation is an individual choice. However, this choice process is not straightforward.

It is useful to consider how radon risk is communicated; typically, health education approaches assume that people will act rationally in response to health threat information: once you tell people that there is a threat, they will be motivated to test to see if they personally are at risk from the threat, they will test, and then they will act to remediate if the test indicates a threat. Efforts are directed to ensure that the information is presented clearly and in appropriate media so that people are exposed to the message. However, the process of translating information received into the relevant behaviours is quite complex. A number of stages can be identified, and Table 1 presents a very simplified breakdown of the processes required for information to result in behaviour.

There are numerous steps involved for the individual in moving from the stimulus (Information that radon may be a threat) to outcomes (testing and remediation of one's home). The steps in the table represent a mixture of content (e.g., the information), processing of the information (which is determined by the individual's perceptions, knowledge, and motivation), and the actual behaviour. The desired behaviours of testing and remediation only occur after all of steps have occurred.

The admittedly simplified idealised sequence is predicated on an assumption that we will respond to health threat information in a rational manner, i.e., that an informed individual tends to behave in the best interests of her/his health (Maziak and Ward 2009). Such an assumption underpins much health awareness campaigns, which judge the success of the risk communication in terms of what the exposed population know about the radon risk. However, knowledge alone does not always influence people's behaviour: radon knowledge is not associated with actually doing a home radon test (Sandman and Eblen 1994). As will be illustrated in this chapter, the "rational actor" assumption gains little support from decades of psychological research on health risk perception and risk-reducing behaviours. People can respond to health threat information in a sub-rational manner, and that such responses reflect powerful unconscious and deliberate psychological processes. In order to better understand the failures of individuals to act to assess and then remediate against the threat from radon, the psychology of risk perception needs to be considered.

Table 1. Steps required for action to occur after radon information programme

1. You are exposed to the information in your environment
2. You attend to the information
3. You understand the information.
4. You believe that there is a threat to you: the information must be perceived as being credible.
5. You understand the threat.
6. You perceive it as a possible risk: the threat may affect you (you are susceptible) and it may have very negative health consequences for you (it is severe).
7. You believe that the threat level can be assessed (i.e., test for radon)
8. You know how to get the threat level assessed.
9. You want to get the threat level assessed.
10. You are confident that you can get the threat level assessed
11. **You act to get the threat level assessed: test.**
12. You understand the result of the test
13. You believe the result of the test
14. You perceive that you are at risk (You are susceptible to a severe negative outcome).
15. You want to reduce this risk.
16. You know how to reduce this risk.
17. You are confident that you can act to reduce the risk
18. **You act to reduce the risk: remediate.**
19. You act to confirm that the risk has been managed: re-test.

Radon exposure accounts for about 9% of deaths from lung cancer and about 2% of all deaths from cancer in Europe (Darby et al. 2005). Given the significant problems arising from radon in houses, radon is a public health issue that needs to be managed. However, the nature of radon and its actual threat level minimize an urgency to act accordingly. Radon is a colorless, odorless, and tasteless gas; consequently, there are no sensory cues to alert people to it as an environmental risk. Research has consistently demonstrated that such cues typically help motivate behavior (McKenzie-Mohr 2000). As radon does not seem to cause any immediate visible health effects and given the absence of other sensory cues the risk is, in essence, out of mind. The experience of the radon risk is benign as people live with the risk, sometimes for many decades, without experiencing any side effects or symptoms. In addition, the effect of the risk is far removed from the initial exposure: for example, lung cancer will develop decades later and, as there are no early symptoms to act as cues, it is easy to delay action. Another challenging feature regarding radon is that it is a naturally occurring risk; in general, evidence indicated that we perceive technological threats to be more risky than natural threats (Sandman 1987). It is arguable that if the lung cancer rate caused by the natural process of radon emission was associated with a manufactured process (e.g., overhead pylon, underground pipes), there would be widespread outrage and immediate calls for governmental action.

In: Nova Excerpts: Earth Sciences
Editor: Dominik Lucas
ISBN: 978-1-53619-038-0

Chapter 12

INDOOR RADON GAS AND ASSESSMENT OF HEALTH RISK*

Mahshid Shahbazi Sehrani[1,†]PhD, Mohsen Mirmohammadi[2], PhD and Siamak Boudaghpour[1], PhD

[1]Department of Civil Engineering,
Khajeh Nasir Toosi University of Technology, Tehran, Iran
[2]Department of Environment, University of Tehran, Tehran, Iran

ABSTRACT

In recent years, the materials utilized and the resources mobilized in technological developments have become the main source of environmental contamination and as a result, health problems. Today, cancer is the most common disease which has got high death rates in all over the world. Lung cancer is one of the most prevalent one. Radon(Rn-222), a natural radioactive gas, is one of the dangerous and fatal indoor air pollutants. Radon is the major cause of lung cancer after smoking. High levels of radon concentration in dwellings and other indoors make serious

* The full version of this chapter is published in *Radon: Detection, Exposure and Control*, edited by Ramesh Chandra Tiwari, published by Nova Science Publishers, Inc., 2020.

† Corresponding Author's Email: sehrani.1991@ut.ac.ir.

and irremediable health problems, although it can hardly be dangerous outdoors. In this chapter, we will introduce radon gas, its origin, mechanism, factors affecting its concentration and methods of measurement. Moreover, we will explain how to reduce indoor radon gas levels. The newest method of health risk assessment from indoor radon will be introduced as well. Finally, the conclusion of some researches will be discussed.

Keywords: air pollution, indoor, lung cancer, measurement, radon, risk assessment.

INTRODUCTION

Radon is an odorless, colorless and tasteless gas. Its dew point equals -62°C and its density is 8 times greater than air. It is a natural gas released from the breakdown of uranium in soils, rocks and groundwater. Radon exposure usually occurs primarily indoors, where the gas will accumulate and convert to high levels [1]. This is a gas with 27 isotopes which are from Rn-200 to Rn-226. These isotopes are not stable and can change to be stable after a series of decay. Only Rn-219, Rn-220 and Rn-222 are naturally occuring of all mentioned isotopes [2]. 1.2 mSv out of 2.4 mSv (safe average dose of annual radiation that a person receives from natural sources) is related to radon [3]. All products of radon decay such as Po-218, Pb-214, Po-214, and Bi-214 are solids. The half-life of these products are 3.05, 0.00016, 26.8 and 19.7 minutes respectively. When radon is breathed, it decays in the lungs through alpha emission and causes some damages. Alpha particles then strike the lung tissues and will result in lung cancer after the long term. The Potential danger of Po-214 and Pb-214 is 20 times greater than Rn-222 [4]. Radon is known as the second cause of lung cancer and the first cause of cancer mortality in the world after tobacco smoking [5-7] leading to be deemed as a class 1 (A) human carcinogen by the IARC (International Agency for Research on Cancer) [8]. The EPA considers Class A carcinogens as pollutants with adequate human data indicating the chemical causes cancer in people. About 90% of lung cancers occur among ever smokers [9]. As well, it is estimated that 86% of radon-related lung cancer deaths are in current and former smokers [10, 11]. It is noticeable that 16% of all lung cancer deaths are related to this gas. Radon is more dangerous for ever smokers which hereinafter is called ES than never smokers which hereinafter is called NS [5, 10, 12, 13]; so that the danger of lung cancer infection is 25 times greater for ES exposed to excess radon concentration than NS [14]. It is highly important to consider that the majority of lung cancers occur among humans exposed to tobacco smoke [5, 15]. According to studies, radon causes 21000 lung cancer deaths per year in the USA. Also, 3200 lung cancer deaths are attributed to radon in Canada every year [16]. Recent studies implemented in Europe, North of USA and Asia assert that the notable number of lung cancers occur because of radon exposure [17-19]. While maximum radon level recommended by EPA (Environment Protection Agency) and WHO (World Health Organization) is about 148 and 100 Bq/m^3 (Becquerels per cubic meter) respectively, Health Canada action level for radon gas is 200 Bq/m^3) [16]; so that it will cause minimum health risks [20]. Bq/m^3 is a measurement of radiation.

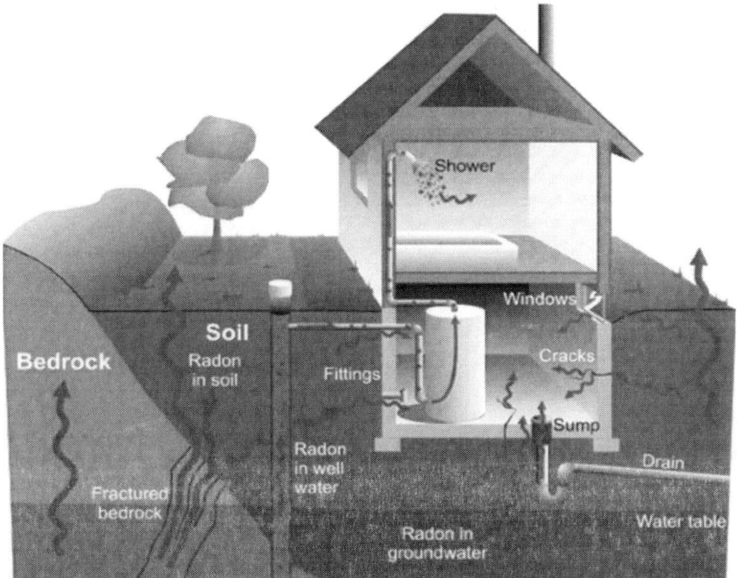

Figure 1. Radon entry points.

How radon can enter an indoor? It is shown in Figure 1. Radon in indoors initially came to attention in 1984. A person named Stanley Watras was an engineer who worked indirectly with radioactive materials at the Limerick nuclear power plant. Once, when he was coming back home, he set off the monitor at a radiation-detection section. Then, he continued to do that for 2 weeks to find the source of radioactive contamination. It was found that the concentration was not related to a work-related exposure, but high radon levels were found at his home. Testing of radon concentration at homes began in 1985 when this experience was published in newspapers. The concentration of radon at that home was measured 4,400 picocuries of radon. The relative risk of his (and his family's) radon exposure was calculated to be equivalent to smoking 135 packs of cigarettes daily. Mr. Watras started a new way to search in radon and reduce radon exposure.

In this chapter, we will find some information about radon in different aspects.

In: Nova Excerpts: Earth Sciences
Editor: Dominik Lucas
ISBN: 978-1-53619-038-0

Chapter 13

A PROPOSAL FOR ACTUATION WITH MEASUREMENTS OF RADON ACTIVITY CONCENTRATIONS IN WORKING PLACES*

A. Martín Sánchez*[†], *M. J. Nuevo and J. de la Torre Pérez
Department of Physics, University of Extremadura, Badajoz, Spain

ABSTRACT

Surveys on radon activity concentration in working spaces are commonly performed by using track detectors. However, these measurements present high variability depending on the time of day and the characteristics of each site. For that reason, only the time spent by employees in their workplaces should be measured by using active detectors. Unfortunately, the usage of many of these devices simultaneously is expensive. In this manuscript, we propose a new strategy in order to optimize concentration measurements at minimum costs.

---

* The full version of this chapter is published in *Radon: Detection, Exposure and Control*, edited by Ramesh Chandra Tiwari, published by Nova Science Publishers, Inc., 2020.

† Corresponding Author's Email: ams@unex.es.

Keywords: Indoor radon activity concentration, radon in working places, track detectors, continuous monitoring, radon surveys

INTRODUCTION

In order to prevent harmful effects on humans, indoor radon activity concentration is controlled by regulating its levels (ICRP 1993; IAEA 2003; WHO 2009; EURATOM 2013). In general, large surveys with a great number of samples are measured indoors by exposing passive devices -such as track detectors- to radon activity concentrations for a given period of time (for example, three months). By following this procedure, annual average concentrations are estimated by taking into consideration the mean value measured throughout the exposure time. The obtained value is then contrasted with those limits given by the corresponding health authority. When necessary, some mitigation actions are carried out to reduce radon activity concentration levels (Coskeran et al. 2009; Veloso et al. 2012). A recent report by the International Commission on Radiation Units & Measurements has warned about the great variations of radon activity concentrations, not only spatial but also temporal, which must be carefully measured and interpreted (ICRU 2016).

Considering that a typical employee spends one third of a working day in their workplace, measurements of indoor radon activity concentrations are also necessary there. For this reason, some studies were carried out in those type of sites (Martín Sánchez et al. 2012). Generally, track detectors are used when a great number of population is surveyed with a large scale of samples. Nevertheless, working space measurements with those detectors may show concentration values above the safety limits, thus requiring mitigation actions when it is the case. Other considerations must be taken into account in these places under study. A thorough analysis by using continuous monitor detectors shows great variations of radon activity concentration within twenty-four hours. Therefore, only mean values measured along the timetable of a regular workday –eight hours- should be considered, since this is really the only time spent by employees in their working places. Significant differences between both values for the concentration have been found in some cases (Martín Sánchez and Nuevo 2017). Consequently, radon activity concentrations should be monitored in deep detail in these situations. For this reason, passive track detectors -just providing average values- can be only considered as a primary indicator. That is why the laboratory has to face a budgetary problem when it comes to choosing between continuous radon monitors (considerably more expensive) versus track detectors.

In: Nova Excerpts: Earth Sciences
Editor: Dominik Lucas
ISBN: 978-1-53619-038-0

Chapter 14

ROLE OF RADON AS A PRECURSOR TO EARTHQUAKES: AN APPRAISAL*

Rajib Biswas†
Department of Physics, Tezpur University, Tezpur, Assam, Inda

ABSTRACT

Radon monitoring has recently gained extensive attention among seismologists. It is now widely reported as a precursory signal before the occurrence of a seismic event. The enhanced estimation of radon in soil gas is basically attributed to the strain changes which arise within the earth's surface. In this chapter, the role of radon as a precursor is elaborated, accompanied by the anomalies which come across during observation. Additionally, the recent studies done in the Himalayan Belt is also reviewed along with future perspectives in radon estimation.

Keywords: radon, precursor, emanation, earthquake

---

* The full version of this chapter is published in *Radon: Detection, Exposure and Control*, edited by Ramesh Chandra Tiwari, published by Nova Science Publishers, Inc., 2020.

† Corresponding Author's Email: rajib@tezu.ernet.in, rajivb27@gmail.com.

INTRODUCTION

In general, precursor signifies something that happened or existed before another thing. Going by this definition, we refer seismic precursor to trends/developments which, in general, precede the occurrence of a seismic event. They can be considered as the forerunner of earthquakes. These precursors involve different physiological changes such as deformation or fracture in ground, rise in seawater level, changes in oceanic tides, foreshocks, anomalies in b-values, radon content, water table fluctuations, magneto-telluric currents and many more. Earthquake prediction phenomena basically entail these observations. They all contribute to the estimation of fundamental parameters of earthqukes which include origin time and magnitude of the impending seismic event. There may be plenty of precursory seismic signals. Yet, they possess considerable drawbacks. One of the main problems lies with the effective separation of signals from inherent noise. It is of the general notion that the signal is always accompanied by noise. Hence, the main hurdle is to distinguish requisite signals from noise of the precursors. Most importantly, we can not gain any useful information just on the basis of a single precursor. It has to be an integral approach engaging various presurssors, albeit under the efficient control of a robust prediction algorithm. Above all, proper evaluation of precursory phenomena directly calls for better comprehension of the physical processes which cause these precursssors. In order to perceive these precursors, one has to resort to physical model of varied attributes. In general, the models are grouped in two classes based on these attributes. The first one deals with the prediction of fault slip behavior without detailing any variations in material properties surrounding the fault. On the contrary, the second one dwells upon relations based on bulk, thereby predicting changes in material properties encircling the fault. In such cases, the first type includes nucleation and lithospheric loading models which need special mention. Similarly, the second category inlcudes dilatancy model as the main basis.

Among the aforementioned precursors, radon estimation has been widely considered as one of the vital parameters. There have been several documentations of monitoring anomalies in emissions of gases like radon, helium, hydrogen, CO_2, etc. However, among researchers, radon emerges to be the favorite one; thanks to easy detection procedure. In this chapter, the observation and analysis of radon monitoring will be highlighted. It will also be endeavored to give a brief outline of the difficuties in the interpretation of data and the relevant foreplay of parameters, accompanying anomalous radon emanations. Further, the recent studies across the Himalayan Belt are also reviewed along with future perspectives.

In: Nova Excerpts: Earth Sciences
Editor: Dominik Lucas
ISBN: 978-1-53619-038-0

Chapter 15

NATURAL RADIOACTIVITY SURVEY ON ROCKS FOUND ON JAMMU – KASHMIR NATIONAL HIGHWAY, HIMALAYAS, INDIA*

***Ajay Kumar*[1,†] *and Sumit Sharma*[2]**
[1]Post Graduate Department of Physics, DAV College, Amritsar, Punjab, India
[2]Department of Applied Sciences, Swami Sarvanand Group of Institutes, Dinanagar, Punjab, India

ABSTRACT

The two major natural sources through which human society is exposed to radiation externally are terrestrial gamma rays and cosmic rays. Terrestrial gamma rays are derivative of natural radio-nuclides that belongs to ^{238}U, ^{232}Th, and ^{40}K series and found in the earth's crust. These radio-nuclides (^{238}U, ^{232}Th, and ^{40}K) are found in soils depending on the distribution of these radio-nuclides in rocks from which they emanate and on the processes at which the soils are concentrated.In this investigation, natural radioactive contents in rocks of Jammu & Kashmir were measured

* The full version of this chapter is published in *Radon: Detection, Exposure and Control*, edited by Ramesh Chandra Tiwari, published by Nova Science Publishers, Inc., 2020.

† Corresponding Author's Email: ajay782@rediffmail.com.

using gamma-ray spectroscopy. NaI(Tl) detector was used to estimate the concentrations of some radio-isotopes present in 15 samples of various rocks collected from Jammu – Srinagar National Highway, India. The average activity concentrations of ^{226}Ra, ^{232}Th, and ^{40}K for the rocks were found to be 28.30 ± 8.34, 42.02 ± 16.40, and 441 ± 68 Bq/kg respectively. Radiological hazard indices such as absorbed dose rate, radium equivalent activity, external Hazard Index (Hex), internal Hazard Index (Hin), Annual Effective Dose (mSv/y), Gamma activity Index (Iγ) and Alpha Index (Iα) were also determined. The mean value of radium equivalent obtained in this study is less than that of the International reference value of 370 Bq/kg in rocks given by OECD (1979). Data obtained indicated that rocks of the studied region did not pose any significant radiological health hazards to the local population.

Keywords: rocks, gamma spectrometer, radionuclide content, radioactivity, radiation hazards

1. INTRODUCTION

Radio-nuclides existing in the environment are either natural or artificial. Uranium and thorium series and ^{40}K nuclides are the most widely dispersed in the lithosphere. The gamma radiations emitted from these primordial radio-nuclides acts as the main external source of irradiation to the human body. The content of the natural radio-nuclides in the soil rely predominantly upon the geographical and geological condition of the soil (UNSCEAR 2008), and to the content in potassium (K), uranium (U), and thorium (Th) of the rock from which the soils emanate in each area. Besides naturally occurring radio-nuclides, there are also artificial radio-nuclides have been released into the environment by different processes. The ^{137}Cs is one of them and is believed to be deposited by atmospheric weapon tests, medical applications, industrial activities and remains in biota and living organisms because of its relatively long physical half-life (Belivermis et al. 2008). Its chemical and eco-physiological similarity with potassium, abundance among fission products, long physical half-life (30.14 years) and relatively high volatility makes radio-cesium one of the most important man-made radio-nuclides released into the environment. Human beings have been exposed to a wide range of ionizing radiations from these naturally occurring radioactive materials (NORMs) and artificial radio-nuclides (UNSCEAR 2011), In order to determine whether the people of the region live in a safe and healthy environment in terms of the natural radio-nuclides, the concentration of radio nuclides in the soil/rocks and the effects of radiation on human health must be known.

The study of natural radioactivity in the soil is important due to their radiological effect and also they serve as an excellent biochemical and geochemical traces (Najam et al. 2011). The estimation of health effects due to exposure of ionizing radiation from natural sources needs awareness of its distribution in the environment. The level of presence of natural radioactivity in soil and water is significant principally due to two radiological effects. The first is the exposure to radiation from radio-elements in diet and inhaled radon and its progenies which emanate from soil, sediment and building materials results in irradiation of the lungs by alpha-emitting short-lived progeny of radon and thoron.The second effect is the external irradiation of the body by gamma rays emitted from primordial radio-nuclides present in the area (UNSCEAR 2000). The long-term exposures to internal and external sources have serious health effects such as chronic lung cancer and leukemia (Sharma et al. 2018, 2018a). Therefore, the assessment of natural radioactivity in the soil is of a great interest for various workers all over the world, which led to the worldwide national surveys in the last two decades (Kumar et al. 2018, Turhan et al. 2012, Celika et al. 2009, Hambye et al. 2002).

The objectives of this present study are to determine activity concentration of long-lived natural radio-nuclides e.g.,^{226}Ra, ^{232}Th, and ^{40}K in rock samples collected from the Himalayan region of Jammu & Kashmir, India using gamma ray spectroscopy. The various radiological hazards i.e., radium equivalent activity, criteria formula, internal and external health effects, gamma index, absorbed dose rate, and annual effective dose equivalent, respectively have been calculated for the health related threat. The result obtained from this study will serve as radioactivity database for the area and will also be relevant in the radiological mapping of the area.

In: Nova Excerpts: Earth Sciences
Editor: Dominik Lucas
ISBN: 978-1-53619-038-0

Chapter 16

SEASONAL VARIATION ANALYSIS OF SOIL RADON DATA*

Hari Prasad Jaishi[1], Sanjay Singh[2] and Ramesh Chandra Tiwari[2,†]

[1]Department of Physics, University of Science & Technology Meghalaya, Ri-Bhoi, Meghalaya, India

[2]Department of Physics, Mizoram University, Aizawl, Mizoram, India

ABSTRACT

This study reports the seasonal variation of soil radon data collected over two different locations viz. Aizawl and Serchhip district in Mizoram, India. The meteorological parameters like rainfall, temperature, relative humidity and pressure were regressed with soil radon data to investigate the effect of these parameters on the measured radon data.In Serchhip, the soil radon measurement was carried out from July 2011 to May 2013 whereas in Aizawl the measurement was carried out from March 2013 to May 2016. The integrated soil radon measurement in Serchhip was carried out after every 15 days and at weekly interval in Aizawl. To check the

* The full version of this chapter is published in *Radon: Detection, Exposure and Control*, edited by Ramesh Chandra Tiwari, published by Nova Science Publishers, Inc., 2020.

† Corresponding Author's Email: ramesh_mzu@rediffmail.com.

seasonal variability of the measured soil radon concentration, the data collected during the entire period of observation has been divided seasonally. It was observed that the soil radon concentration vary seasonally at both the locations. Higher values were obtained during summer season and lower values were observed during spring and winter seasons.

Keywords: soil radon, seasonal variation, correlation, meteorological parameters

INTRODUCTION

Radon is a radioactive noble gas present in all the rocks of the earth. In 1990, a German scientist, Fredrick Dorn found that radium (Ra-226) emanates a gas.He named the gas as "radium emanation." Later in 1908, Ramsey and Whytlow-Grey called it "Niton" (the shining one). In 1918, Schmidt introduced the name "Radon" which has become the universal name of this element (George, 2008). Atomic number of radon is 86 and it has thirty seven known isotopes (Neriet al., 2011). The majority of the radio isotopes have short half-lives.Out of thirty seven known isotopes, only three of them are naturally occurring namely: Rn-222 (radon), Rn-220 (thoron) and Rn-219 (actinon). Rn-222 belongs to the U-238 decay chain produced by the α-decay from its mother nuclide Ra-226; Rn-222 has a mean half-life of 3.823 days. Rn-220 is produced by the α-decay of Ra-224 which belongs to the decay chain of Th-232 and has a mean half-life of 56 sec (Nazaroff, 1992). Rn-219 is derived from the decay series of U-235 and has a very short mean half-life of 3.96 sec.Of all the three isotopes, the half-life of Rn-222 is long enough to undergo a diffusive transport through thick soil layers or other materials. Therefore, it is considered to be most useful in geochemical surveys. The half-life of Rn-220 is relatively short, so it is useful only in those areas where advective transport of soil-gas is particularly high or in areas where there are rocks containing significant amounts of thorium-rich mineral phases. Rn-219 decays in the soil before it can reach the atmosphere due to its very short half-life. So, the role of Rn-219 in geochemical exploration is negligible. Moreover, being a gas Rn-222 near the ground surface has a great tendency to escape into the atmosphere. Besides, radon concentration in ground surface increases with depth which depends on the soils properties and moisture content (Jonsson, 1995; Kristianson and Malmqvist, 1984).

Migration of Radon and Thoron in Soil

Radium and thorium in soil are the main source of radon and thoron in the atmosphere. Even though the activity exhalation rate for thoron is much higher, about 100 times (Harley, 1973) than that for radon, the total global activity inventory in the atmosphere is much less, about 100 times less than that for radon, primarily because of the shorter half-life of thoron. The physical process of radon and thoron release from the structure of the earth is not yet fully understood. Only a fraction of radon or thoron atoms generated in the soil escapes from the solid grains and makes its way into the pore of the medium. The ratio of the amount of radon escaping from the solid matrix and that produced by the radioactive decay is known as the *Emanation coefficient* or *Emanating power* of rocks. Tanner (1980) hypothesized a physical mechanism explaining radon emanation from soil by recoil effect as shown in Figure 1. He suggested that radon atoms which terminate their recoil path into the liquid or gas phase may escape from the crystalline matrix (solid grains).

Those atoms which are located at a distance from the edge of grain which is greater than the recoil length, or those which terminate their path inside the adjacent grain, remain trapped within solid phase and thus not available for convective transport. The recoil energy of radon and thoron atoms are 86 and 103 KeV respectively (Bossus, 1984), four or five order of magnitude greater than the ordinary bond energies which is enough to move the atom from the original position (Stoker and Kruger, 1975). The newly formed atom travel from its site of generation until its energy is transferred to the host material. The distance travelled depends on the density and composition of the material (Nazaroff, 1992). The range of recoil of radon is 0.02–0.07 μm for common minerals, 0.1μm for water and 63 μm in air. The range of recoiling thoron in air is 83 μm (Tanner, 1980).

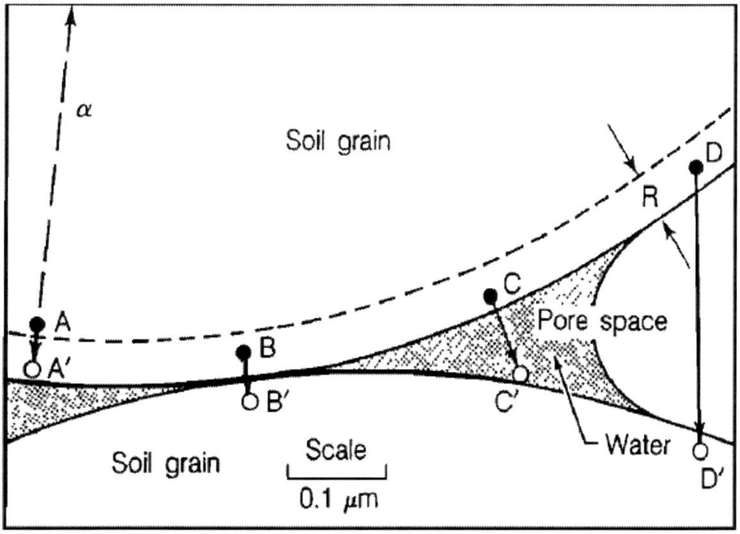

Figure 1. Schematic sketch of the possible emanation scenarios in a solid-water-air system. Radium atom, indicated by solid circles, decay, producing an alpha particle and a recoiling radon atom which may end its recoil at the point indicated by the open circle; R =recoil range; case A-A') ^{222}Rn remains inside the grain; case B-B') ^{222}Rn reaches the adjacent grain; case C-C') ^{222}Rn atom reaches the water filled pore space, possibly reaching the air filled pore space by diffusion; case D-D') recoil energy is large, ^{222}Rn crosses the air filled pore space and penetrates into the adjacent grain (after Tanner, 1980).

The emanation coefficient can be strongly influenced by the moisture content in the soil (Strandenet al., 1984), increasing with soil moisture, up to saturation in the normal range of soil moisture content. In a very dry solid the release of radon or thoron is reduced by re-adsorption of radon and thoron atoms on surfaces in the pores and fractures in the solid. If the solid is slightly moist, the release is enhanced up to certain moisture content, above which it decreases again because of the lower diffusion rate in water-filled pores (Megumi and Mamuro, 1974). In addition to the soil moisture effect, grain size and temperature dependence has also been observed. Radon emanation increases with decreasing grain size (Markkanen and Arvela, 1992; Morawska and Jeffris, 1994). This is explained by the recoil theory and the fact that the smaller the grain size, the larger the surface area which is available for Ra-226 adsorption (Megumi and Mamuro, 1974). A rise in temperature also causes an increase in emanation coefficient. Strandenet al., (1984) found that radon exhalation rate for a soil sample increases by 55% with the increase in temperature from 5 to 50°C. They suggested that this rise in temperature may be due to the reduced adsorption of radon.

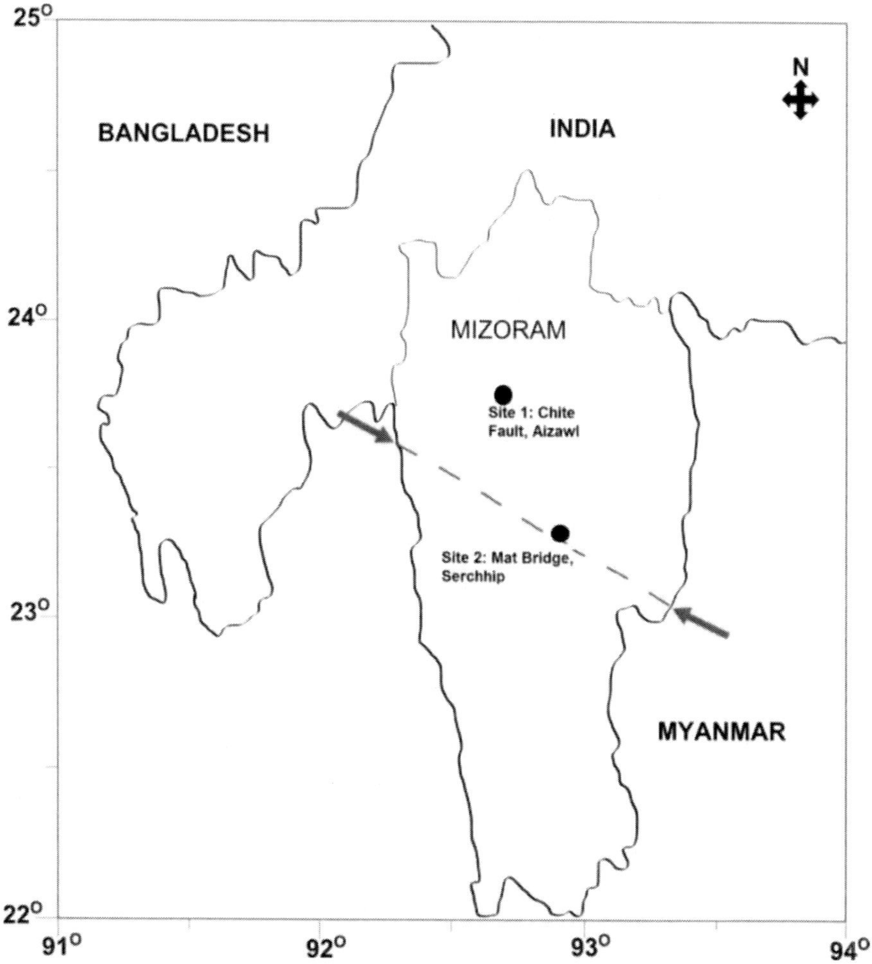

Figure 2. Geographical location of the study area. The red dashed line shows Mat Fault, the most prominent fault in Mizoram.

The anomalousbehavior of soil radon and thoron data with seismic events for the study area has been reported (Jaishiet al., 2013; 2014a; 2014b; 2014c; 2015; Singh et al., 2014; 2016; 2017). However this study describes the comparative study on seasonal variation of soil radon and thoron data. The emanation of soil radon and thoron concentration depends upon several environmental parameters such as air temperature, soil moisture, relative humidity and air pressure. In this study we perform the seasonal analysis of soil radon and thoron over two different regions in Mizoram India. The location map of the study area is shown in Figure 2.

In: Nova Excerpts: Earth Sciences
Editor: Dominik Lucas
ISBN: 978-1-53619-038-0

Chapter 17

ANALYSIS OF METEOROLOGICAL INFLUENCE ON EXHALATION OF ^{222}RN AND ^{220}RN GASES AT MAT FAULT*

T. Thuamthansanga[1], R. C. Tiwari[1,†], B. K. Sahoo[2] and D. Datta[2]

[1]Department of Physics, Mizoram University, Aizawl, Mizoram, India
[2]Radiological Physics and Advisory Division,
Bhabha Atomic Research Centre, Mumbai, Maharashtra, India

ABSTRACT

In-situ online ^{222}Rn, ^{220}Rn, air temperature and relative humidity data were measured between November, 2017 and April, 2018 with a frequency of once in a month at Mat fault (23.3°N, 92.9°E), Mizoram, India. The influence of the two 15 minute's average hourly recorded meteorological factors on exhalation of 5 minutes average hourly recorded ^{222}Rn/^{220}Rn gases were reported. It was observed that for the hourly recorded data, ^{222}Rn and ^{220}Rn gases diurnally varies in direct correlation with air temperature but in reverse correlation with relative humidity. The influence

* The full version of this chapter is published in *Radon: Detection, Exposure and Control*, edited by Ramesh Chandra Tiwari, published by Nova Science Publishers, Inc., 2020.

† Corresponding Author's Email: ramesh_mzu@rediffmail.com.

of air temperature and relative humidity on 12 hours average $^{222}Rn/^{220}Rn$ exhalation rate were also analysed and compared with the previous findings made at the same fault. A ZnS(Ag) based scintillation counter (Model: SMARTRnDuo, BARC, India) equipped with capacitor type digital sensor (Model: DHT22/AM2302) was deployed for measuring the ^{222}Rn, ^{220}Rn, air temperature and relative humidity data.

Keywords: radon, ZnS(Ag) scintillation, *in-situ* online measurement, meteorological parameters, correlation

INTRODUCTION

Radon is a radioactive noble gas having three naturally occurring isotopes viz. ^{222}Rn ($T_{1/2}$=3.8 days), ^{220}Rn (55.6 s) and ^{219}Rn (3.6 s). Due to its extremely small half-life ^{219}Rn often gets neglected in most of radon-related studies, as well as in the present study. ^{222}Rn and ^{220}Rn were originated from their respective parent nuclei ^{238}U and ^{232}Th; emanated by the process of recoil effect in the earth crust. From there they got transported to the earth surface by the process of diffusion and convection for exhalation into the atmosphere. The diffusion length of ^{222}Rn and ^{220}Rn in typical soil are 1 m and 1 cm, respectively, the diffusion equation for the isotope pair in a typical soil is given by equation (1) (Sahoo et al., 2016)

$$\frac{\partial C_p}{\partial t} = S - \nabla . F_p - \lambda C_p \tag{1}$$

where C_p is the radon activity per volume of the pore space, S is the radon activity released into a unit volume of the pore space per unit time, F_p is the activity of radon crossing per unit pore area per unit time and λ the decay constant.

Radon studies had been carried out from the past till date in order to understand its activity level in the atmosphere due to its potential health risk (Duckworth et al., 2002; Nazaroff and Nero, 1988; Sahoo et al., 2007), as a geochemical exploration tracer (Baskaran, 2016; Dyck, 1968; Morelli et al.,

2006) and as a precursor to seismic activity (Jaishi et al., 2013; Singh et al., 2016; Ulomov and Mavashev, 1967). Emission of radon gas from the soil was directly influenced by meteorological factors (Klusman and Webster, 1981; Singh et al., 1988). In case of radon studies for a precursor to geophysical phenomena, identification of radon anomaly from meteorological origin and geophysical origin is of vital importance in order to avoid a false prediction. In the present study, the authors report the influence of meteorological parameters (air temperature and relative humidity) on the emission of ^{222}Rn and ^{220}Rn gases from the soil at one of the prominent fault (Mat fault) in Mizoram state, India (Figure 1). A ZnS(Ag) scintillation based counter (Gaware et al., 2011, 2011a, 2013) equipped with a capacitive type digital sensor was deployed for measuring *in-situ* online ^{222}Rn/^{220}Rn concentrations, surface relative humidity and air temperature. Upon analysing the hourly (approx.) recorded data between 5:00 and 17:00 hours of IST (Indian Standard Time) a direct correlation was observed between ^{222}Rn/^{220}Rn data and air temperature while a reverse correlation with relative humidity. But from the 12 hours average data obtained between 5:00 and 17:00 of IST, ^{222}Rn shows a reverse and direct relationship with air temperature and relative humidity, respectively while ^{220}Rn maintains its same fashion with the two meteorological parameters as that of the hourly recorded data. In general, it can be concluded that diurnally ^{222}Rn varies directly with air temperature and reversely with relative humidity during the study period while the 12 hours (approx. daily) emission rate of ^{222}Rn might decrease with air temperature but increases with relative humidity. ^{220}Rn on the other hand, shows positive and negative correlation with air temperature and humidity respectively for the hourly recorded data and in the 12 hours average data. The *in-situ* online generated data will serve as a future reference for radon-related studies carried out in the region.

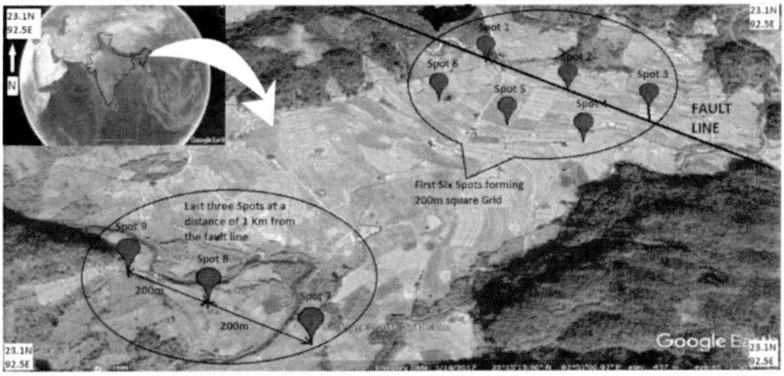

Figure 1. Location Map of the study area showing formation of measuring spots at Mat fault.

In: Nova Excerpts: Earth Sciences
Editor: Dominik Lucas
ISBN: 978-1-53619-038-0

Chapter 18

RADON TESTING AND MITIGATION IN SCHOOLS OF SHILLONG CITY, MEGHALAYA*

Devesh Walia[1,†], Deveshwori Maibam[2], Yubaraj Sharma[3] and Atul Saxena[4]

[1]Department of Environmental Studies, North-Eastern Hill University, Shillong, Meghalaya
[2]Department of Physics, Don Bosco College of Teacher Education, Tura Meghalaya
[3]Department of Physics, Don Bosco College, Tura, Meghalaya
[4]Department of Physics, North-Eastern Hill University, Shillong, Meghalaya

ABSTRACT

Radon is a naturally occurring radioactive inert gas and a member in the natural decay of the primordial element Uranium, whose isotopes are omnipresent in varying trace amounts in soils and rocks. It is a major

* The full version of this chapter is published in *Radon: Detection, Exposure and Control*, edited by Ramesh Chandra Tiwari, published by Nova Science Publishers, Inc., 2020.

† Corresponding Author's Email: wadevesh@gmail.com.

source of radiation dose to humans accouting for more half the total natural exposure.Most of the exposure to the general population happens inside closed spaces like buildings, where radon can accumulate to significantly higher levels than the outdoor. Children with smaller lung volumes, higher breathing rates and rapidly growing cells are particularly vulnerable to radiation damage and since a considerable amount of time is spent by them in schools, the present study aims to assess the risk by estimating Potential Alpha Energy Concentration(PAEC) and radon activity concentration in schools in Shillong and to estimate the contribution to the Annual Effective Dose (AED) to the students and teachers of the schools. The survey is conducted in seven schools of Shillong city using SSNTDs, LR-115 Type2 detectors in membrane with filter mode and bare modes in two seasons, for winter 2017 and summer 2018. PAEC obtained from bare exposed detectors rangesfrom 0.035-0.099 WL and 0.010-0.068 WL respectively. The Radon activity concentrations obtained from detectors exposed in membrane with filter mode ranges from 126.0-650.0 Bq.m^{-3} and 31.9-197.3 Bq.m^{-3} respectively. The AED values are found to be less than the average annual effective dose limit for members of the public prescribed by UNSCEAR, 1993 (1.2 mSv.y^{-1}).

Keywords: PAEC, Schools, LR-115 Type2, Radon Testing, Radon Mitigation

INTRODUCTION

The annual average exposure to radiation from various natural sources is estimated to be about 2.4 mSv; radon contributes majorly to this, primarily through inhalation exposure (~53%) and also less ostentatiously through other mediums (United Nations Scientific Committee on the Effects of Atomic Radiation 2008). Radon progenies either directly inhaled or through the in vivo decay of inhaled radon, tend to deposit along the walls of the various airways of the bronchial tree and provide the main pathway for radiation exposure of the lungs. These exposures are mostly produced by alpha particles emitted by the radionuclides, which irradiate the secretory and basal cells of the upper airways and are responsible for reported lung cancer risks (Marsh et al. 2008). Potential Alpha Energy Concentration (PAEC) is a measure of the human exposure to radon and its decay products in the air that is breathed in and expressed in units of WL (Working Level) or Joule m^{-3}(United Nations Scientific Committee on the Effects of Atomic Radiation 2000; Srivastava et al. 1996). Indoors, radon can accumulate to harmful levels whereas in an outdoor environment, radon is quickly diluted and is of no concern from a radiological point of view (Faheem and Matiullah 2008). An indoor environment like schools can be a significant source of radon exposure for children and the staff working there (Vaupotič, Hunyadi, and Baradács 2001; Bilban and Vaupoti 2001; Venoso et al. 2009). Children (students) have smaller lung volumes, higher breathing rates and rapidly growing cells, hence, they are at considerably higher risk from elevated radon concentrations.

The present study attempts to assess the radiological risk due to radon in seven schools located at Shillong, the capital city of the North-eastern Indian state of Meghalaya (Fig. 1). The seven schools for the study were selected considering accessibility, location and cooperation from the authorities. The tabulation of the schools' GPS locations is given in Table 1.

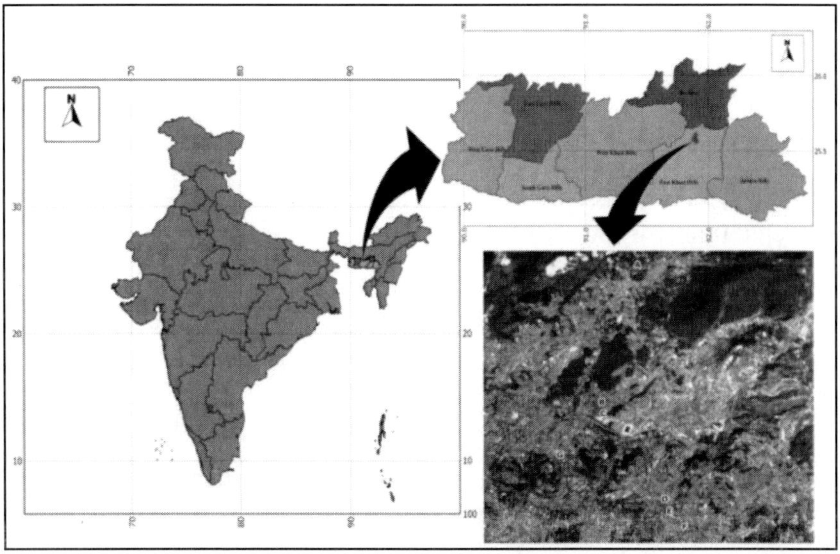

A: Kendriya Vidyalaya –NEHU, B: Buddha Vidyalaya, C: St. Xavier Sec, D: Laitumkhrah Presbyterian Hr Sec, E: Brookside Adventist Sec.Niketan Upper Sec., F: Nongthymmai Nepali Higher Sec., G: R.B Anoop Chand Hindi Sec.

Figure 1. A map showing the location of the study sites.

Table 1. Location of the surveyed schools

Code Name of the Site	Name of the Schools	Latitude	Longitude
A	Kendriya Vidyalaya – NEHU	N 25^0 36'25.1''	E 091^0 53'43.1''
B	Buddha Vidyalaya Niketan Upper Secondary	N 25^0 35'02.1''	E 091^0 53'21.6''
C	St. Xavier Secondary	N $25^0$34'58.4''	E $091^0$53'22.4''
D	Laitumkhrah Presbyterian Hr. Secondary	N 25^0 34'03.8''	E 091^0 53'59''
E	Brookside Adventist Secondary	N 25^0 33'59.6''	E $091^0$54'02.2''
F	Nongthymmai Nepali Higher Secondary	N 25^0 33'47.0''	E 091^0 54'11.1''
G	R.B. Anoop Chand Hindi Secondary	N 25^0 34'30.6''	E 091^0 52'56.2''

A major motivation for the present study stems from the fact that the study area has many reported cases of cancer and is located over the Shillong Plateau with basement of Granite Gneiss, which normally has more concentration of Uranium and Radium, and nearness to the one of the largest uranium deposit in India and complex geology with prominent faults/thrusts including Dawki fault which may provide conduit to the radon produced at depth to reach the surface and increases the probability of elevated radon levels (Walia *et al.*2010).

In: Nova Excerpts: Earth Sciences
Editor: Dominik Lucas
ISBN: 978-1-53619-038-0

Chapter 19

CALIBRATION FACTOR FOR LR-115 TYPE II DETECTOR FOR SOIL-GAS RADON MEASUREMENT*

Devesh Walia[1,†], Atul Saxena[2], Deveshwori Maibam[3] and Yubaraj Sharma[4]

[1]Department of Environmental Studies, North-Eastern Hill University, Shillong, Meghalaya

[2]Department of Physics, North-Eastern Hill University, Shillong, Meghalaya

[3]Department of Physics, Don Bosco College of Teacher Education, Tura Meghalaya

[4]Department of Physics, Don Bosco College, Tura, Meghalaya

ABSTRACT

Calibration of the detectors is an important factor that limits the accuracy of radon measurements using Nuclear track detectors. The widely

---

* The full version of this chapter is published in *Radon: Detection, Exposure and Control*, edited by Ramesh Chandra Tiwari, published by Nova Science Publishers, Inc., 2020.

† Corresponding Author's Email: wadevesh@gmail.com.

used calibration chamber method for estimating the calibration factor might have limitations in the case of soil-gas radon, which have much more severe field conditions than indoors. The present study estimates calibration factor for integrated radon concentration measurements in soil-gas environment with LR-115 Type II detectors (bare mode) and is standardized using BARASOL MC2. The calibration factor obtained is 0.0383 tracks.cm^{-2}.day^{-1}.(Bq.m^{-3})$^{-1}$ with a standard deviation of 0.00276.

Keywords: Calibration factor, LR115 (type II) detectors, SSNTD, Radon measurements, soil – gas environment, dwellings

INTRODUCTION

In last three decades many new nuclear track detectors such as LR-115, CR-39, ZnP-glass, SR-86 etc. were identified and developed with enhanced sensitivity and resolution. After appropriate calibration, it is possible to use these detectors for measuring radon concentration in soil gas. These detectors are useful in determining time integrated indoor as well as soil-gas radon concentrations in-situ in a very simple and cheap process. Hence, these detectors are important reconnaissance tools for any further detailed work using other instruments with higher accuracy, and most suitable for large scale deployment of time integrated measurement of radon concentration levels (Jönsson, 1997, Jönsson, 1999 Durrani, 2008, and Walia, et al., 2010).

Among the various detectors, the LR-115 Type II detector film is popular among researchers due to its sensitivity and ease in etching and microscopic study. The detector is a 100μm thick polyester substrate pasted with a 12μm thick layer of red coloured cellulose nitrate which allows most of the incident alpha particles to penetrate through it forming well-etched perforated holes (tracks) (Dwaikat, et al., 2007, Dwivedi, 1995). Due to its wider application in the measurement of indoor radon; LR-115 has been calibrated mostly for use in the indoor environments (Hakam, et al. 1995, Ramu, et al., 1988, Srivastava, *et al.*, 1996 and Eappen&Mayya, 2004,). Calibration exercises for LR-115 detectors for the measurement of soil–gas radon have been lacking whereas issues with this approach have been reported (Rector, 1991).

For better accuracy in the measurement of soil-gas radon concentration using LR-115 as detector, it is essential to calibrate the LR-115 detectors under conditions which simulate in-situ conditions, taking into consideration the variations of environmental parameters due to differing geographical and geological locations. In the current paper, we describe a calibration exercise and present the calibration factor for LR-115 Type II detectors in bare mode for soil-gas radon.

In: Nova Excerpts: Earth Sciences
Editor: Dominik Lucas
ISBN: 978-1-53619-038-0

Chapter 20

ATMOSPHERIC CORROSION IN THE CANARY ISLANDS*

J. J. Santana[1,†], PhD, V. F. Mena-González[1], MSc, H. C. Vasconcelos[2,3], PhD, S. González[4,5], PhD and R. M. Souto[3], PhD

[1]Department of Process Engineering, University of Las Palmas de Gran Canaria, Las Palmas de Gran Canaria, Gran Canaria, Canary Islands, Spain
[2]Faculty of Sciences and Technology, Azores University, Portugal
[3]Centre of Physics and Technological Research (CEFITEC), FCT/*UNL*, Portugal, Centre of Biotechnology of Azores (CBA), Portugal
[4]Department of Chemistry, University of La Laguna, La Laguna, Tenerife, Canary Islands, Spain
[5]Institute of Material Science and Nanotechnology, University of La Laguna, La Laguna, Tenerife, Canary Islands, Spain

---

* The full version of this chapter is published in *Archipelagos: Types, Characteristics and Conservation*, edited by Mabel Perez and Juana Mills, published by Nova Science Publishers, Inc., 2018.

† Corresponding Author Email: juan.santana@ulpgc.es.

Abstract

Corrosion is a process that occurs naturally and irreversibly. It is defined as the deterioration of a material (usually a metal) due to its interaction with the surrounding environment. Economic losses related to corrosive processes account for about 5% of the Gross National Product (GNP) of Western countries, China and India. Only in the United States, the costs related to metal corrosion are estimated at app. $1,000 annual *per capita*, and responsible for a substantial portion of this cost is atmospheric corrosion. Within the various types of corrosive processes, atmospheric corrosion attracts great interest since it affects the entirety of metal materials exposed to air, including urban and industrial infrastructures. It is a widely studied process, and there are national and international standardized norms for its characterization classifying the aggressiveness of a given environment for a specific metal type (usually those most widely used by the industry). However, major discrepancies between the results of direct exposure and those obtained from the application of standardized norms and predictive models occur for fragmented geographical areas, as in the case of archipelagos. In the Canary Islands, due to their characteristic orography and the regime of winds that affects the archipelago, a multitude of microclimates is generated in a very small geographical area. This variability of atmospheric environments in different islands means, even for the same metal, that the extent of the atmospheric attack may vary substantially between two locations separated only a few miles away. This fact imposes the need for the development of specific prediction models to characterize the corrosion rate for different metals. Investigations carried out in the Canary Islands allowed the characterization of the atmospheric corrosion for metals of wide industrial use (namely mild steel, galvanized steel, zinc, copper, and aluminium), and to obtain maps of atmospheric corrosivity for all the islands. Specific mathematical models have been developed for the prediction of the corrosion rates for these metals exposed on the various islands that make up the archipelago.

Keywords: atmospheric corrosion, corrosion category, Canary Islands, weight loss, metals, predictive models

1. Introduction

Corrosion is both the cause and the effect of the chemical interaction of a material with the surrounding environment. In the case of metals, this process of deterioration is called metallic corrosion.

The corrosion of metallic materials is a major financial and ecological plague since repairs due to corrosion account for about 5% of the Gross National Product (GNP) in Western Countries, China, and India [1]. The consequences are a loss of resources, energy, and the release of toxic substances to the environment due to the oxidation reactions. The most important type of corrosion process is of electrochemical nature. In this process, anodic and cathodic reactions are involved. In order to prevent a corrosion process, one or both redox reactions must be hindered.

Metallic corrosion is a natural process that is thermodynamically favored. As any physicochemical process, the corrosion rate depends on the metal and the environmental variables. When a metal, alloy or metal coating is exposed to the action of the atmosphere, the occurring spontaneous process is called *atmospheric corrosion*. Its rate, as in any other corrosion process, depends on a large number of variables such as the nature of the metal, environmental factors (rain, fog, dew, relative humidity, wind, temperature...), and the amounts of contaminants (SO_2, NO_x, chloride, etc.).

Atmospheric corrosion is the most important cause of deterioration of metals given the enormous number of structures and metal parts of all kinds (e.g., bridges, electrical transmission towers, roofs, gates,...) that operate exposed to the atmosphere. Thus, the availability of atmospheric corrosivity maps facilitate the task of selecting the right material for a particular application in a given location based on the local atmospheric aggresivity.

Numerous studies on atmospheric corrosion have been performed and published [2, 3], including the development of international standard norms aimed at normalizing this type of studies. One of the limitations presented by these works is that they refer to continental areas with minor geographical and seasonal climatic variabilities. By contrast, the analysis of the atmospheric corrosivity in an archipelagic region like the Canary Islands presents many complexities. We have observed important limitations in the existing international standard norms [4] and a great variability of results compared to those found in continental areas.

In: Nova Excerpts: Earth Sciences
Editor: Dominik Lucas
ISBN: 978-1-53619-038-0

Chapter 21

PERSPECTIVES OF ATMOSPHERIC SCIENCES IN THE AZORES: HISTORY OF METEOROLOGY AND CLIMATE CHANGE*

Maria Gabriela Meirelles[1,†]
and Fernanda Rosário Silva Carvalho[2,‡]
[1]Department of Physics, Chemistry and Engineering,
University of Azores, Azores, Portugal
[2]Afonso Chaves Observatory,
Portuguese Institute of the Sea and the Atmosphere,
Azores, Portugal

ABSTRACT

Knowledge of the processes by which each area of science constructs its foundations and establishes filters is a permanent link that, to a certain extent, explains and contextualizes its historical milestones and the degree

---

* The full version of this chapter is published in *Archipelagos: Types, Characteristics and Conservation*, edited by Mabel Perez and Juana Mills, published by Nova Science Publishers, Inc., 2018.

† Corresponding Author Email: maria.gf.meirelles@uac.pt.
‡ Corresponding Author Email: fernanda.carvalho@ipma.pt.

of progress in the present. Thus, the history of meteorology begins in the early era from the time when man understood that weather and climate conditions were essential for planning present and future activities. Thus, the history of meteorology begins in the early era from the time when man understood that weather and climate conditions were essential for planning present and future activities. The acquisition of knowledge about time has been improved over the years, taking advantage of the technological advance.

Due to its geographical location, the Azores archipelago is by nature one of the perfect places to observe and study the characteristics of atmospheric processes. The evolution of Meteorology took place in different phases - the empirical one during the years of 1860 to 1920, transition phase took place between the years of 1920 to 1950 and the scientific phase began in 1950 and continues until the present time. The science of meteorology in the Azores was not alien to this dynamics, and during the empirical and transitional era, four observatories were built on different islands of the archipelago. However, since 1817 there are initially discontinuous records in the time and space of the meteorological parameters, pressure, temperature and wind observed at the surface. The decades of the 50s and 60s were fertile in advancing the knowledge of the Earth system, particularly the atmosphere and the simulation of its behavior through models. It should be noted the contribution of the Department of Meteorology of MIT, where names such as Victor Paul Starr, Edward Norton Lorentz, Jule Charney and Norman Phillips, among others, laid the groundwork for what constitutes the current view of the general circulation of the atmosphere and Weather Forecast. With Starr and his observational method, the meteorology left the keplerian phase and entered in the Newtonian phase.

The archipelago of the Azores has an important historical heritage in the area of meteorology and contributes with the records of the meteorological observations made in the islands for future climatic projections. From the moment these projections are known, we are in a position to assess the archipelago's possible vulnerability to climate change, which is one of the main environmental factors responsible for the negative impacts on biodiversity. Thus, it becomes possible to propose measures of adaptation and conservation of the archipelago in face of identified vulnerabilities, contributing to the development and implementation of a regional strategy for adaptation to climate change.

The consequences of climate change for the Azores are also discussed. The remote conditions of the islands of the Azores allowed the atmospheric monitoring of greenhouse gases without the influence of local sources of pollution. The increase of the carbon dioxide in the atmosphere is also observed in the Azores, as well as the temperature of the air near the surface. Projections of the CMIP5 for the Azores show an increase in the average annual temperature in the Azores region that can reach 4K by the year 2100 and a decrease in average annual precipitation by about 600 mm.

Keywords: meteorological history, climate change and climate models

INTRODUCTION

Knowing the climate of a particular location is fundamental to the planning of human activities. (Durschmied, 2004), how much blood, sweat, and tears have ever been shed because of changes in climate? Atmospheric phenomena have changed the battles, made history, charged and continue to charge dearly for lives and sufferings. We can mention, for example, the typhoon that decimated the fleet that would invade Japan in the thirteenth century, the cold that froze the German panzers in Second World War There are battles decided more by the climate than by courage or strategy. The extreme phenomena happened, happens and will continue to happen.

Throughout its journey humanity has sent large amounts of greenhouse gases into the atmosphere. However, it was from the nineteenth century that, due to the intensive use of transport and the increase of industrial activities, man began to contribute, to a large extent to a possible climate change, referred in (Somerville and Hassol, 2011). Savante Arrenhius, 1896, used the CO_2 concentration at the time, which was around 300 ppm (parts per million) and based on calculations, he knew that a doubling of CO_2 concentration would imply an increase in average temperature by about 6°C.

The Intergovernmental Panel on Climate Change (IPCC), which was established in 1988, is a scientific organization, politically oriented and belonging to the United Nations. Its aims are to compile and disseminate the work done by scientists around the world about climate change. The causes, effects and risks of climate change for the environment and for humanity are referred. Also, the adaptation and mitigation measures are mentioned on Assessment Reports published by IPCC.

After we transcribe the citation made by Intergovernmental Panel on Climate Change (IPCC) in its 4th report published in 2007, "The climate is like a train that has begun its march and that, during the next centuries, no one will be able to stop the route."

In the biennium 2013-2014, the 5th IPCC Report was published, in which there is information that until the end of the century, the planet will heat up to 4.8 degrees Celsius, which will cause the sea level rise to 82 centimeters. Drought periods and heat waves will also multiply, and intense rainfall, which is likely to cause an unprecedented wave of homeless. In addition, the Arctic region will heat up faster than the global average, and warming will be greater on land than on the ocean. There is a high risk of substantial impacts on terrestrial and aquatic ecosystems as a result of climate change, causing especially negative consequences for biodiversity and ecosystem services. The rate of warming, the magnitudes of acidification and warming of the oceans and rising sea levels, increase the risk of harmful effects on ecosystems and human systems. Understanding how climate phenomena work is to understand the interaction between the sun, the atmosphere, oceans, ice, relief and vegetation, thus linking science with the history of civilizations.

Historical meteorological parameters are very important to understand the climate of the past and to calibrate the models that forecast the future climate based on scenarios. General information about the network of weather stations in a country or region generally appears to be trivial and, for this reason, no record is kept in the required details. However, for historical metadata, all the details are relevant. From the late nineteenth century, the Meteorological and Hydrological Services of the different countries began to understand the need to harmonize all observations and measurements and to regulate the publication of data. Recommendations and resolutions concerning meteorological stations, observations and instruments of measurement were discussed and approved at the International Meteorological Organization (IMO) Conference of Directors held in 1879. Since then, the regulations and practices used in the processes of observation and measurement have been modified and adapted to new circumstances, taking into account the technological advances.

The history of a meteorological station or of a network of meteorological stations has great importance in the interpretation of the differences detected in the historical series. In section 4 we will refer the case found in the series of temperatures of the city of Horta, island of Faial, for the period 1902-1932.

The archipelago of the Azores lies in the Atlantic Ocean between latitudes 36° 45'N and 39° 43'N and longitudes 24° 45'W and 31° 17'W, distant from the eastern end about 1400 km west coast of mainland Portugal. The nine islands of the archipelago of the Azores extend for a range of about 600 km in length, according to three distinct groups:

- the Western group includes the islands of Flores and Corvo,
- the Central Group, the Terceira, Graciosa, São Jorge, Pico and Faial Islands,
- The Eastern Group integrates the islands of São Miguel and Santa Maria.

In: Nova Excerpts: Earth Sciences
Editor: Dominik Lucas
ISBN: 978-1-53619-038-0

Chapter 22

BIOGLASS PREPARED VIA A MINERAL WATER-ASSISTED SOL-GEL ROUTE*

H. C. Vasconcelos[1,2,3,†], R. Coutinho[1,4], P. Freire[4], A. C. Farias[1], M. C. Gonçalves[5] and A. Silva Pinto[1]

[1]Faculdade de Ciências e Tecnologia,
Universidade dos Açores, Portugal
[2]Centro de Física e Investigação Tecnológica (CEFITEC),
FCT/*UNL*, Portugal
[3]Centro de Biotecnologia dos Açores (CBA), Portugal
[4]Instituto de Vulcanologia e Avaliação de Riscos (IVAR)
Universidade dos Açores, Portugal
[5]Centro de Química Estrutural (CQE), Instituto Superior Técnico (IST)
Universidade de Lisboa, Portugal

* The full version of this chapter is published in *Archipelagos: Types, Characteristics and Conservation*, edited by Mabel Perez and Juana Mills, published by Nova Science Publishers, Inc., 2018.

† Corresponding Author Email: helena.cs.vasconcelos@uac.pt.

ABSTRACT

Furnas Volcano (São Miguel Island), in the Azores archipelago, became popular in the early nineteenth century for its mineral water sources used for medicinal purposes. Furnas has several sources of thermal waters all highly ionized with pH values ranging from highly acidic to alkaline. These qualities make the Furnas mineral water ideal for the sol-gel synthesis of ceramics and glass with the added benefits of being both a reactant and catalyst for this chemical process.

In the 1960s, Hench first introduced bioactive glass to be used in bone implants. Due to frequent corrosion exhibited by the metals of the first generation of biomaterials, Bioglass 45S5® and other bioactive glass began to be developed and soon attracted the attention of many researchers. This is mostly due to its ability to establish more efficient biological and mechanical connections with bone tissue by forming a layer of carbonated hydroxyapatite (HCA) on the surface.

The structure and composition of glass can host a wide range of metallic ions which dictate its bioactivity. When liberated these ions can mediate bone regeneration.

Because in the sol-gel process the precursor materials are in water, the incorporation of metal ions is generally easier providing for a wide range of compositions and direct doping ions for therapeutic purposes.

This chapter discusses a new paradigm in the sol-gel synthesis, particularly in the synthesis of bioactive glass. In this new approach the deionized water is replaced (in the hydrolysis step) by highly ionized and naturally acidified thermal water collected from springs located in Furnas Volcano. Given the potential impact of these waters on the structure and bioactivity of sol-gel bioactive glass, special emphasis was given to their genesis, physicochemical properties and ionic diversity.

To conclude we present a case study on the synthesis and characterization of bioactive glass (45S5, 58S and 77S systems) using several types of thermal waters with the respective chemistry shown in a Piper diagram.

Keywords: bioactive glass, sol-gel, volcanic waters, Azores Archipelago

In: Nova Excerpts: Earth Sciences
Editor: Dominik Lucas
ISBN: 978-1-53619-038-0

Chapter 23

THE RED SEA: A NEWBORN OCEAN*

Laura Pinarelli, PhD†
Institute of Geosciences and Earth Resources, Section of Florence, National Research Council (CNR), Florence, Italy

ABSTRACT

During much of the Phanerozoic until the Oligocene, the African and Arabian plates were part of a single shield. The history of its dislocation via the Red Sea opening starts in the late Oligocene (~24 Ma), when the Arabian plate separated from that of Nubia through a rupture of the Precambrian shield and rotated counter-clockwise. After advanced continental rifting, oceanic spreading began in the Pliocene (~5 Ma) and continues to the present day, as demonstrated by the intense volcanic and seismic activity in the area during the last 10,000 years.

Seafloor magnetic anomalies and normal faulting earthquakes, typically associated with spreading ridges, occur in the Red Sea south of 20°N, where a deep axial trough is present. To the north, only few isolated anomalies are identified, and the sea is floored by rotated fault blocks. South of 16°N the axial trough narrows and some groups of islands emerge,

---

* The full version of this chapter is published *in Red Sea: Historical Significance, Properties and Economic Importance*, edited by Delbert Robbins, published by Nova Science Publishers, Inc., 2018.

† Corresponding Author Email: lapina@igg.cnr.it.

the last of which was formed in 2013 in the Zubair Archipelago. The northern area is interpreted as continental crust being stretched and faulted with only minor volcanic activity, whereas the southern area has evolved to the point that new lithosphere is being produced at a spreading axis. This situation has been viewed as transient until the extension in the northern area evolves into seafloor spreading.

The southern deep axial trough is the focus of intense submarine volcanic activity. The geochemistry of erupted magmas conveys valuable information about the underlying mantle. They are low potassium tholeiitic basalts with K2O < 0.2% and TiO2 < 2.0%. Their Sr and Nd isotope ratios are ≈ 0.703 and ≈ 0.513, respectively. These geochemical features are typical of mid ocean ridge basalts (MORB), and allow the Red Sea to be regarded as a newborn ocean, whose development likely resembles that of the Atlantic Ocean in its early stages, and which may one day become the sixth ocean of the planet. It is interesting to note that the erupted lavas are characterized by geochemical variations along the Red Sea. The basalts of the central part exhibit geochemical parameters that resemble those of MORB. Both north and south, the basalts display higher contents of incompatible elements, higher Sr and lower Nd isotope ratios, as well as higher Pb isotopes ($^{206}Pb/^{204}Pb$ up to 19.608), than those recorded in the central part, suggesting a contribution by "enriched" mantle components.

Keywords: Red Sea, basalts, geochemistry, isotopes

1. Introduction

Since the 1960s the Plate Tectonic scientific revolution demonstrated that the Earth is subjected to cyclical processes, whereby the continents assemble in a single supercontinent, and then gradually fragment and disperse again. A fundamental step inside this cycle of continental breakup, dispersal, and reassembly, is the splitting of a continent and the growth of a new ocean. A major breakthrough in understanding the way the modern Earth works has been highlighted by the studies of the ocean floor: the Joint Oceanographic Institutions Deep Earth Sampling (JOIDES). This project began in 1969, continued with the Deep Sea Drilling Project (DSDP), and since 1976 it has been replaced by the International Phase of Ocean Drilling (IPOD) project. These projects have produced more than 500 boreholes in the seafloor worldwide, and the results achieved have been as outstanding as the plate-tectonic theory itself. They confirmed that the oceanic crust is everywhere younger than about 200 million years, and that the age of the oceanic crust calculated from the magnetic anomalies is close to the age of the overlying oceanic sediments estimated by micropaleontology.

Reconstructions of the geological past suggest that the continental masses from roughly 250 to 150 Ma were assembled in a single Supercontinent, called Pangea. The Red Sea is one of the components of a continent scale rift system located along a divergent plate boundary that, since the Oligocene, divide the African, Somalian, and Arabian plates (Figure 1), called Afro-Arabian Rift System (Baker 1970; Khan 1975). Besides the Red Sea, it includes the Gulf of Aden and the Ethiopian Rift, which together form a triple junction located at Afar (Gass 1970; McKenzie et al., 1970; Burke and Dewey 1974; Le Pichon and Francheteau 1978). Considering the evolutionary history of the planet, the Red Sea can be regarded as a fairly recent formation. It likely started forming ≈ 24 million years ago, when tectonic movements caused the separation of the African plate from the Eurasian one. The seismic activity that has always characterized the floor of this sea, causing an expansion of the basin between 10 and 16 mm/yr (Figure 1), is still present today.

In: Nova Excerpts: Earth Sciences
Editor: Dominik Lucas
ISBN: 978-1-53619-038-0

Chapter 24

MINERAL RESOURCES IN THE RED SEA/GULF OF ADEN: POLYMETALLIC-BARITE MINERALIZATIONS AT WADI MASILAH BASIN (YEMEN)*

Laura Pinarelli, PhD[1,†] and Mohamed A. Mattash[2]

[1]Institute of Geosciences and Earth Resources, Section of Florence, National Research Council (CNR), Florence, Italy

[2]Ministry of Oil and Mineral Resources, Geological Survey and Minerals Resources Board, Sana'a, Yemen

ABSTRACT

Yemen is characterized by a considerable number of metallic and non-metallic raw materials, including base metals, precious metals, and dimension stones. The Wadi Masilah Basin, in the southeasternmost part of Yemen, hosts very promising base metal and barite prospects. This terrain has undergone a complex geological and structural history,

---

* The full version of this chapter is published *in Red Sea: Historical Significance, Properties and Economic Importance*, edited by Delbert Robbins, published by Nova Science Publishers, Inc., 2018.

† Corresponding Author Email: lapina@igg.cnr.it.

including oblique divergence (transtension) related to the opening phases of the greater Red Sea – Gulf of Aden rift system, with NW-SE and NE-SW major fault zones.

Barite occurs either singly or in association with galena, hemimorphite, willemite, zincite, smithsonite, descloizite, celestine, quartz, calcite, dolomite, iron and manganese oxides, and gypsum, connected mainly to the Cretaceous carbonates (Saa'r Formation). The mineralisations crop out as either vein fillings or stratiform type. $BaSO_4$ contents between 95 and 98% and specific gravity values up to 4.37 g/cm^3 are indicative of high quality ores.

Sulfide mineralisations consist of sphalerite, galena and pyrite. They occur in association with barite veins or form fracture-filling in limestones and silicified dolostones. Available data indicate several lead-zinc ore bodies at various depths, some of which are considerably large with dimensions attaining 60 m x 30 m x 50 m, and tonnages reaching thousands of tons.

Nonsulfide minerals comprise hemimorphite, willemite, smithsonite, hydrozincite, and zincite, as well as cerussite, hydrocerussite, anglesite, wulfenite, and rarely pyromorphite. Zinc contents in the mineralised veins range from low (1-8 wt.%) to high grades (30-50 wt.%). The above mentioned minerals occur in a variety of structures, such as: shear zones, fractures, vugs, cavities, brecciated veins, and open spaces of the vein barite sidewalls, and connected mainly to the dolostone sequence.

Vanadium ores mostly occur as descloizite and rarely vanadinite, in association with barite, calcitized veins and brecciated zones, and dolomitic limestones, with vanadium contents up to 12.30 wt.%. These prospects contain one of the most remarkable occurrences of descloizite known, and besides lead-zinc, descloizite is probably the mineral resource for which the Wadi Masilah Basin is most promising.

Mn and Fe ores are also present, whose mineralogy consists of hollandite, romanechite, cryptomelane, pyrolusite, hematite, goethite and limonite. They include several discordant irregular veins. The results of geochemical analyses reveal high quality Mn ores, with an average MnO content of 64%.

The orebodies in all prospects consist mainly of fracture-filling in limestones, dolostones, and silicified limestones. They originated from both hydrothermal processes and the action of slightly acidic groundwater. The prospecting area is structurally controlled and located in a NW-SE major fault zone and its parallel lineaments. Tectonic uplift and intensive fracturing systems, along with favorable climatic conditions, gave rise to the development of supergene Zn, Pb, V, Mn and Fe mineralisations.

Keywords: Yemen, Wadi Masilah Basin, polymetallic mineralisations, barite

1. INTRODUCTION

Non-renewable raw materials are fundamental for industrialised economies. In spite of the rapid growth of renewable energy, industrialised countries still rely heavily on non-renewable resources. Moreover, many emerging economies are now entering the phase of strong industrialisation that requires significant amounts of energy, mineral resources, metals and other raw materials. Mineral deposits and metals are among the most important materials required by almost all countries for construction, production of tools and equipment, for wiring and manufacturing of automobile body parts. Even though mineral deposits are comprised mostly of naturally occurring elements, their extraction and processing is very expensive. Nowadays, there is an increasing interest in the supergene polymetallic deposits, particularly after successfully developing of a number of procedures that facilitate their processing and concentration (e.g., Falagàn et al., 2017).

Yemen is characterized by remarkable deposits of metallic and non-metallic raw materials, including base metals, precious metals, and dimension stones. A variety of ancient mines were found across the country, of gold, silver, copper, iron, lead, etc., which were exploited by the ancient Yemenis to extract their metals and oxides, used to produce metal objects, cosmetics, pigments, ceramics, ornaments. The last Century has witnessed several intensive explorations, including geological, geochemical, and geophysical surveys, and thematic mapping. Such activities were systematically undertaken by Yemeni, European, and North American Institutions and Societies. Recently, a new Mining Law has been introduced to encourage foreign and local companies to invest in the minerals, petroleum, and gas sectors, as well as in the geothermal energy resources.

This chapter reports a synthesis of many years of geological and geochemical investigations on the Wadi Masilah Basin, in particular at Alkohl1, Alkohl2, and Alqalanah areas, which contain the first barite occurrences in Yemen, as well as supergene non-sulfide zinc minerals, and vanadium rich zones (Mattash et al., 2007a and b; 2008; 2009; 2010). These outcrops are amongst the metallic-nonmetallic occurrences of Yemen that are already partially exploited.

In: Nova Excerpts: Earth Sciences
Editor: Dominik Lucas
ISBN: 978-1-53619-038-0

Chapter 25

CURRENT STATUS OF THE SUDANESE RED SEA FISHERIES RESOURCE MANAGEMENT*

Omyma Karrar, PhD
Arab Open University, Kuwait City, Kuwait

ABSTRACT

The Sudanese Red Sea is considered to be unique in many aspects. It is still pristine with rich and diversified ecosystems and high biodiversity in addition to other potential resources of high economic value. Among these resources is a rich fisheries reserve dominated by different assemblages of an ecologically and commercially important stock of invertebrates, demersal finfish and pelagic fish. The national fisheries standing stock, excluding the shared stock, remains largely under-exploited with the exception of Mother of Pearl (*Pinctada margeritifera*) and Trochus (*Tectus dentatus*) which are efficiently utilized. The fisheries resource if fully exploited will greatly contribute to the food security of the region and provide sources of income to the local communities. Official statements and reviews including international and regional reports by the

* The full version of this chapter is published *in Red Sea: Historical Significance, Properties and Economic Importance*, edited by Delbert Robbins, published by Nova Science Publishers, Inc., 2018.

Food and Agriculture Organization of the United Nations ((FAO), 1989), United Kingdom Overseas Development Administration ((ODA), 1996), Regional Organization for the Conservation of the Environment of the Red Sea and Gulf of Aden ((PERSEGA), 1998) and Sudan National Report (2013) have identified a number of challenges that the Sudanese Marine Fisheries Sector is facing in order to increase the catch yield.

Keywords: Sudanese Red Sea, Marine fisheries sector, standing stock, food security, catch yield

About this Chapter

The first section of this chapter will provide general information about the study area and outline the background of the issue. Next, the findings and key issues will be discussed including the constraints and challenges of the Sudanese Red Sea fisheries resource management. Finally, future directions including actions by local policy makers to address existing challenges both in terms of activities for improving the yield and measures to ensure sustainable use of the resource are provided.

Introduction

Background Information

The Sudan, officially known as The Republic of Sudan was the largest country in the continent of Africa until the separation of South Sudan in 2011. It is now the third largest country in Africa and 16th largest in the world, with an area of 1,861,500 km^2. Sudan lies between latitudes 8° and 23°N, within the North Eastern Part of Africa and is bordered by the Red Sea to the east, and seven countries. The bordering countries and their relative location are as follows: Egypt to the north, Eritrea and Ethiopia to the east, South Sudan to the south, the Central African Republic to the southwest, Chad to the west and Libya to the northwest. The country's terrain is generally made up of flat plains, interrupted by several mountain ranges, including the Marrah Mountains (3,042 m or 9,980 ft) in the western region, which form the highest point in Sudan, and the Red Sea Hills in the eastern region (Sudan Country Profile, 2017). The River Nile is considered to be the most significant feature in the country's topography, forming a prominent incision in the terrain and dividing the country into eastern and western halves. The River Nile, which flows northwards through Egypt and ultimately to the Mediterranean Sea, consists of the Blue, and White Nile rivers, which meet in Khartoum, the Capital of Sudan.

Sudan's population according to the most recent census reports, is 40,533,550, with an annual growth rate of 2.38% as reported by the United Nations Population Division for the year 2017 (United Nations, 2017). The country is rich in both living and non- living resources. The natural resources include reserves of crude oil, precious metals, significant animal wealth and massive areas of fertile arable land. Agriculture (including fisheries) being the backbone of the country's economy. Agriculture contributes 42 percent to the annual GDP, however, the contribution of the fisheries sector is relatively limited. Sudan's exports are minute and were estimated to be 0.2 million dollars in 2012. The country also relies on imports of fish and fishery products amounting to 5.2 million USD dollars.

Sudan's fisheries sector is known to have a rich resource base acquired from both marine and fresh water reserves. The marine resource includes offshore and inshore waters. The fresh water resource is mainly based on the River Nile and its tributaries, wetlands, lakes and man-made reservoirs. The capture fisheries' production was estimated to be about 34, 000 tons in 2012, 29,000 tons from fresh inland water catches and 5,000 from marine catches. Capture fisheries' activities are centered around the River Nile and its tributaries, seasonal flood plains and four major reservoirs, as well as the territorial waters of Sudan on the Red Sea (Country Brief, 2014).

It is important to note the heavy involvement of international Non-Governmental Organizations (NGOs) and several other agents including UN (United Nations) agents in the Sudan as result of the many years of civil war, existing internal conflicts in addition to a large influx of refugees from neighboring countries. It attempts to help alleviate the difficulties faced by displaced people and the country in general. Therefore they have a strong presence throughout the country. The NGOs and UN agents operating in Sudan include the World Food Program (WFP), the Food and Agriculture Organization of the United Nation (FAO), the United Nations Development Program (UNDP), the United Nations Industrial Development Organizations (UNIDO), the United Nations Children Fund (UNICEF), the United Nations High Commission for Refugees (UNHCR), the United Nations Mine Service (UNMAS), the United Nations Office for the Coordination of Humanitarian Affairs (OCHA), World Bank, and the International Organization for Migration (IOM). Although these organizations are primarily providing humanitarian aid, they contribute significantly to the development of the country through joint developmental projects, funds, technical support, and training, in addition to consultancies and expertise. Examples of the influence and involvement of these organizations in relation to Sudan's agriculture and fisheries resource management will be discussed in this chapter.

Study Area

Red Sea State: Sudan adopts a federal system and consists of 18 States, with the Red Sea State (Al-Baher Al-Ahmer) being one of them. It has an area of 212,800 km^2 and an estimated population of 1,396,000 (Sudan Country Brief, 2017). The Red Sea State is divided into nine administrative areas: Port Sudan- the capital of the state and the main port, Suakin- the second sea port, Gebiet Haya, Sinkat, Gunob Awlieb, Derodieb, Tokar and Ageeg. As the name suggests, the Red Sea borders the state to the east. Inland, mountains run from north to south, interrupted by arid plains. To the northwest of the state is the Nubian Desert. The Siyal Islands are located in the disputed area between Egypt and Sudan in the northeast.

The population of the Red Sea State consists of rural nomads, rural settlers and urban people. The original inhabitants of the state are the Beja tribes, who make up over 65% of the current population, In Sudan, the Beja, who are of Hematic origins, have been considered as indigenous and the largest ethnic tribe in the eastern state. Within the Beja tribes, the Amarar and Besharin sub-groups occupy the northern part of the Red Sea State and are both nomadic pastoralists and fishermen. The Hadandaw and Beni Amer sub-groups in the south Kassala state are agro-pastoralists and fishermen (Saeed, 2004).

The Sudanese Red Sea coast lies within the Red Sea State and extends for 750 km including embayment (Schoreder et al., 1980). The Exclusive Economic Zone is 91,600km^2 (square Kilometers) of which 22,300km^2 are fringing reefs. The major features include a lining of fringing reefs, coastal lagoons, barrier reefs and atolls. Several of these lagoons are fringed by mangroves (Khalil & Krupp, 1994), which along with coral reefs support many fisheries. The productivity of the Red Sea is relatively poor as it is a semi-enclosed water body, however biodiversity and species endemism are quite high. It is estimated that between 10% and 30% of the Red Sea fish are not found anywhere else (Randall, 1992). The potential of the Red Sea resources is fairly high and levels of exploitation are generally low. This fact was noted by member countries bordering the Red Sea, and with a sense of regional responsibility, those countries formed a regional coalition to assess the fisheries resources status and address relevant challenges. The outcome has since been a Strategic Action Plan which was formulated as an immediate need for the Red Sea Countries with vested interest in promoting regional development (PERSGA, 1998).

Capture fresh water fisheries were found to be the main contributors to the total national production, whereas the Marine fisheries contribution was determined to be relatively small (Farah, 2000). Overall, it is frequently reported that the Marine fisheries resource utilization is not enough to satisfy the national demand, and the demand had to be partially met by imports, often resulting in lower average consumption per capita rates as the prices of fish products are generally higher than meat. The country is also dependent on imports of fish and fishery products (USD 5.2 million in 2012) to satisfy the limited per capita fish consumption (about 1.1 kg in 2012). Exports on the other-hand are insignificant and were valued at 2 million USD in 2012 (Country Brief, 2014).

These exports are mainly attributed to marine fisheries products such as finfish, Crustacean, Mollusca, and sea cucumber in addition to limited amounts of processed food (dry and wet salted fish). Some of the Marine Fisheries products are exported regionally to some Arab countries and neighboring African countries. Some products are exported to Europe and Asia.

Figure 1. Map of Sudan States including the Study Area (Al Bahr al Ahmar State – The Red Sea State).

In: Nova Excerpts: Earth Sciences
Editor: Dominik Lucas
ISBN: 978-1-53619-038-0

Chapter 26

DATING THE LAST DEGLACIATION: GEOCHRONOLOGIC METHODS USED TO ASSIGN TEMPORAL CONSTRAINTS TO THE RETREAT OF LATE PLEISTOCENE ICE SHEETS*

Ken Munyikwa†
Centre for Science, Athabasca University, Athabasca, Alberta, Canada

ABSTRACT

Assignment of a temporal scale to deglaciation events is elemental to the understanding of processes involved during and subsequent to the retreat of continental ice sheets. Dating the recession of ice sheets, however, is complicated by the fact that ice sheets cease to exist following deglaciation, leaving behind no tangible remnants of the ice itself that can directly be analyzed. Hence, dating exercises essentially aim to place constraints on ice sheet recession by examining proxy records that resulted

* The full version of this chapter is published *in Deglaciation: Processes, Causes and Consequences*, edited by Delbert Robbins, published by Nova Science Publishers, Inc., 2017.

† Corresponding author: Centre for Science, Athabasca University, 1 University Drive, Athabasca, AB T8N1T3, Canada, Email: kenm@athabascau.ca.

from or are associated with the former existence of ice sheets. A deglaciation timescale enables rates of postglacial geomorphic processes as well as the pace of environmental evolution to be determined. Additionally, such chronologies facilitate the analysis of variations in the distribution of biotic communities over time, in conjunction with changes in climatic conditions and other aspects of the environment. In areas where isostatic rebound occurred following glacial unloading, accurate dating allows geophysical models of viscoelastic responses to be calibrated, including changes in sea level. This chapter provides a review of dating methods that have been used to constrain the retreat of Late Pleistocene ice sheets from formerly glaciated regions following the Last Glacial Maximum, with a focus on North America. Chronological methods explored include cosmogenic-nuclide exposure, luminescence, radiocarbon and varve dating techniques. In each instance, the principles behind the dating technique and the types of material that can be dated are examined. The chapter concludes with a brief outline of a case study conducted in western Canada to constrain the Late Pleistocene retreat of the Laurentide Ice Sheet from the region.

Keywords: deglaciation, dating methods, ice sheet, Late Pleistocene, Quaternary, Laurentide Ice Sheet, ice age

INTRODUCTION

Previous deglaciation events were associated with diverse changes in earth's atmospheric, physical and biotic environments (e.g., Gibbard and Head, 2009; Clark et al. 2012; Cuffrey et al. 2016). A cogent insight into the mechanisms of such events lends an understanding to earth's response mechanisms to climate fluctuations. The ability to reconstruct past deglaciations, however, rests on the capacity to assign a robust temporal scale to observed features and processes. By virtue of its proximity in time, the last deglaciation is the most studied glacial termination to date, with investigations having begun at least 175 years ago (Stroeven et al. 2016). Several geochronologic techniques can be employed to provide absolute age constraints to processes involved in the retreat of Late Pleistocene ice sheets from terrestrial regions (e.g., Jackson et al. 1997, Dyke et al. 2001, 2003; Balco and Schaefer, 2006; Raukas et al. 2010; Ridge et al. 2012; Stroeven et al. 2016; Munyikwa et al. 2017). Assigning a chronology to the recession of ice sheets, however, is complicated by the fact that ice sheets cease to exist following deglaciation, leaving behind no tangible vestige that can directly be analyzed. Hence, geochronologic methods primarily aim to place constraints on ice sheet recession by examining proxy records that resulted from or are associated with the former presence of large ice sheets on continental landscapes.

Observable terrestrial effects of large-scale continental glaciation following ice sheet recession can generally be grouped into three main categories. The first of these comprises erosional features that result from the abrasional effects of ice sheets over bedrock on which the glaciers move. Following deglaciation, such eroded surfaces become exposed to the atmosphere and there are dating methods that can be used to assign temporal constraints for the exposure of the surfaces. Surfaces of weathering products that are produced by glacial processes are also included in this category. Ages obtained from dating the surfaces would be commensurate with minimum age constraints for the deglaciation.

The second group of proxy records consists of depositional materials emplaced by glaciers. The materials include sediments deposited directly by melting glaciers, such as glaciofluvial materials, or glacial sediment reworked by eolian processes during the early postglacial period. Sediments that are released by glaciers, if deposited in water bodies such as lakes as annual layers, can also be used to determine ages that constrain the time of deglaciation. In all instances, ages from the depositional features would serve as minimum age constraints for the timing of glacial retreat.

Since glaciated regions are generally not amenable to most faunal and floral organisms, such terrains are generally barren. Hence, the third and final category of proxy indicators of deglaciation includes biotic communities that colonize formerly glaciated regions as climatic conditions ameliorate. By determining the ages of the earliest organic materials to emerge on the postglacial landscape, it may be possible to establish the earliest time by which the organisms would have been able to flourish following the retreat of ice sheets. Hence, the chronologies would be conservative estimates for the timing of the retreat of the ice sheets.

The importance of assigning constraints to ice sheet retreat cannot be overstated. The dating allows ages of postglacial landforms to be ascertained (e.g., Balco and Schaefer, 2006; Munyikwa et al. 2017). It also enables rates of postglacial geomorphic processes as well as rates of environmental evolution to be determined. In areas where isostatic rebound occurred following glacial unloading, accurate dating allows geophysical models of viscoelastic responses to be calibrated (e.g., Tarasov and Peltier, 2006). Furthermore, environmental variables that are ascertained with an effective temporal resolution enable the leads and lags to be differentiated (e.g., Lowe and Walker, 2015), which would be an important distinction when determining forcing factors.

This chapter provides a review of dating methods that have been used to constrain the retreat of Late Pleistocene ice sheets from formerly glaciated regions following the Last Glacial Maximum (LGM). Focus is placed on terrestrial records and, while emphasis will be on North American ice sheets, with particular attention on the Laurentide Ice Sheet (LIS), the dating methods examined should be globally applicable. Notably, the LIS was the largest Late Pleistocene ice sheet in the Northern Hemisphere. Additionally, though it was also the last of the major ice sheets to terminate, it had a major effect on when the earth system transitioned to Holocene conditions (Ullman et al. 2015). Thus, the ice sheet is of major interest to many researchers. This chapter comprises ten sections. Following the introduction in the first section, the Quaternary chronological framework as well as causes of glaciation and deglaciation during the Pleistocene are examined in the second section. In the third section, features of formally glaciated landscapes that can be dated using geochronologic methods are looked at. The fourth section presents an overview of dating methods that can be used to constrain deglaciation before examining specific methods in subsequent sections. In the fifth section, the use of cosmogenic radionuclides for ascertaining exposure ages of formerly glaciated terrains is explored. The sixth section examines the use of luminescence dating methods to constrain depositional ages of sediments on postglacial landscapes. Age constraints based on radiocarbon dating performed on remains of biological organisms, floral or faunal, that colonize postglacial landscapes are reviewed in the seventh section. In the eighth section, varve chronology as well as other dating methods that could be used to date deglaciation are explored. As an illustration, a case study conducted in western Canada to constrain the retreat of the LIS from the region is briefly discussed in the ninth section. The chapter concludes in the tenth section by examining future trends and methodological advances in dating techniques that may expand prospects in dating deglaciation going forward, including dating glacial retreats beyond the last deglaciation.

In: Nova Excerpts: Earth Sciences
Editor: Dominik Lucas
ISBN: 978-1-53619-038-0

Chapter 27

THE INFLUENCE OF DEGLACIATIONS IN THE MID-LATITUDE EUROPEAN CLIMATE*

***Gloria M. Martin-Garcia*†**
Department of Geology, University of Salamanca, Salamanca, Spain

ABSTRACT

Global climate during the Quaternary has been deeply influenced by glacial-interglacial oscillations. Since the onset of glaciations, the Earth has experienced alternations between warm and stable climatic periods – coinciding with interglacials, and cold and highly variable climatic intervals – coinciding with glacials. In a suborbital timescale, climate oscillations were maximal during glacial onsets and, very especially, during deglaciations. It is very obvious that the extension of ice sheets has a direct influence on high-latitude climate, both in continents and oceans; contrary to this, climate in middle latitudes is not so influenced by global ice volume or the total extension of ice sheets, but by episodes of rapid variation of sea ice coverage. Since early Pleistocene the most pronounced stadial (cold) events in the mid-latitude North Atlantic occurred during deglaciations, both during glacial/interglacial transitions – Terminations –

* The full version of this chapter is published *in Deglaciation: Processes, Causes and Consequences*, edited by Delbert Robbins, published by Nova Science Publishers, Inc., 2017.

† Corresponding author Email: gm.martin@usal.es.

and episodes of lesser ice sheets depletion. These severe cooling events were followed by very rapid, short-lived warming episodes that, when coinciding with Terminations of climatic cycles, initiated the climatic optimum of the following interglacial. Such millennial scale oscillations affected not only sea surface temperature, but also continental climate as far as the eastern Mediterranean basin. Most deglaciation onsets coincided with maxima in Northern Hemisphere summer insolation. Enhanced heat budget facilitated rapid ice sheets collapse and allowed the release of huge amounts of fresh water to the ocean, triggering the reduction or even the shutdown of North Atlantic deep water formation, which resulted in the advection of subpolar waters along the western European margin and the associated stadial. After deglaciation, once the freshwater perturbation stopped, the meridional overturning circulation rapidly resumed and warm surface water recirculated again in the North Atlantic, producing the fast temperature rise and initiation of interglacial climatic optimum across Europe.

Keywords: quaternary, deglaciations, climate change, continental climate, SST, North Atlantic, European climate

ABBREVIATIONS

AF	Arctic Front
AMOC	Atlantic meridional overturning circulation
AzC	Azores Current
AzF	Azores Front
AzH	Azores high-pressure area
HE	Heinrich events
NA	North Atlantic
NAC	North Atlantic Current
NADW	North Atlantic deep water
NAO	North Atlantic oscillation
NAO^+	Positive phase of the NAO
NAO^-	Negative phase of the NAO
ITCZ	Inter-tropical Convergence Zone
ENSO	El Niño/southern oscillation
SST	Sea surface temperature

T Termination.

INTRODUCTION

Climate Variation during the Quaternary

Glacial-Interglacial Oscillation

During the Quaternary period global climate has been characterised by periodic oscillations of advance/retreat of the ice sheets that partly covered the Earth´s continents (e.g., Lisiecki and Raymo, 2007). Each climatic cycle begins with a very warm interval known as interglacial climatic optimum, when ice volume reaches a minimum value. Models suggest that some of these ice volume minima, like during marine isotope stage (MIS) 11 – from ~430 to ~365 ky, were even lower than present day values (Raynaud et al. 2003). The climatic optimum is followed by progressive cooling and increase of ice volume, marked by cool-warm oscillations of lesser amplitude and higher frequency than glacial-interglacial ones, which culminates with the highest extension of ice sheets during the cycle, known as glacial maximum. Each climatic cycle finishes with a rapid deglaciation, known as Termination, which connects with the following interglacial.

During the last million of years these climatic cycles have followed an approximate periodicity of 100 ky with superimposed climatic oscillations of ~41 ky and ~21 ky (e.g., Imbrie et al. 1992; Ruddiman, 2006; Huybers, 2007), all of which can be correlated with the orbital cycles described by Milankovitch (1941). Nevertheless, at a lower time scale, discrepancies exist between climatic records and orbital parameters, which suggest the influence of other factors in modulating the global climatic system. Some of these factors are thought to be the atmospheric concentration of greenhouse gases (Shackleton and Pisias, 1985; Kennett et al. 2000; Ruddiman, 2003, 2006; Toggweiler et al. 2006), oceanic circulation (Broecker et al. 1989; Raymo et al. 1997; McManus et al. 2004; Pisias et al. 2010), ice sheets size, and albedo effect (Imbrie et al. 1993; Denton el al., 2010).

Millennial-Timescale Climate Variation

During middle and late Pleistocene several millennial-scale iceberg surges into the North Atlantic occurred that altered the ocean´s heat conveyor and caused fundamental reorganisations of the ocean-climate system, producing extreme cooling (Heinrich, 1988; Broecker, 1994, Cortijo et al. 1995; Hodell et al. 2008; Channel et al. 2012). These episodes, known as "Heinrich events" (Broecker et al. 1992), had a duration of a few hundred to a thousand years and a recurrence of ~7-10 ky (e.g., Bond and Lotti, 1995). Superimposed on this variability, a series of large shifts in air temperature, up to 15ºC (Huber et al. 2006), in a few decades (Stuiver and Grootes, 2000), termed Dansgaard–Oeschger (D–O) oscillations have been registered in the Greenland ice-core records (e.g., Dansgaard et al. 1993; Grootes et al. 1993); the final event coincided with the Younger Dryas (e.g., Bond et al. 1992). These stadial-interstadial oscillations appear to be grouped in sequences of progressively cooler events, the end of each sequence marked by the deepest cooling; such sequences are known as "Bond cycles" (e.g., Broecker et al. 1990) and are thought to be linked to changes in the mode of overturning circulation in the Atlantic ocean (Broecker et al. 1985).

Most of Heinrich events happened at the end of the colder phase of a Bond cycle and were followed by warming to almost interglacial temperatures (e.g., Bond et al. 1992; Wright and Flower, 2002).

Climate in the Subtropical and Middle Latitude North Atlantic and Western Europe

Although ocean dynamics (especially heat transport) are responsible for controlling sea surface temperature (SST) on interdecadal timescale, interannual variations are mainly driven by atmospheric heat flux changes (Bjerknes, 1964). In this sense, the climate regime over middle latitude North Atlantic and Western Europe is mainly governed by the position and intensity of the Azores high-pressure area (AzH) and its latitudinal motion and intensification through the annual seasonal cycle. During spring/

summer the AzH migrates northward, which results in strong northerly winds, while during fall/winter months the AzH is significantly weaker (Maze et al. 1997; Coelho et al. 2002). In consequence, this region is under the influence of the North Atlantic Oscillation (NAO) anomaly (Hurrell, 1995). This anomaly is measured by the NAO index, which records the difference in atmospheric pressure between the Iceland low and the AzH. The anomaly is positive (NAO^+) when the AzH is reinforced and the difference with the Iceland low is greater. In this scenario, westerly winds become stronger and the Gulf Stream migrates northward, which results in anomalously mild and wet winters for northern Europe, while southern Europe becomes anomalously cold and dry. The NAO also affects the sea ice coverage in the North Atlantic (Deser et al. 2000) and, indirectly, the thermohaline circulation (Dickson et al. 1996). In SW Europe, NAO^+ phases are associated with reinforcement of the seasonal upwelling off Portugal, via strengthening of NE trade winds. In NAO^- conditions the westerlies are weaker, the Gulf Stream is located southward, climate in NW Europe is dry while in the south is wet and off Iberia the Azores Current prevails, inducing warm conditions and the reduction of the upwelling (Dickson et al. 1996).

In tropical regions, climatic conditions are mainly governed by the alternation of rainy (warm) and dry (colder) seasons that trace the seasonal motion of the ITCZ. The southward migration of the ITCZ appears to be linked to cooling on the North Atlantic over various time-scales (Black et al. 1999; Peterson et al. 2000; Schmidt et al. 2004) through weakening of the Atlantic Meridional Overturning Circulation (Schiller et al. 1997).

The climate of southern Europe, specifically the Iberian Peninsula, is also connected with the ENSO through more complex mechanisms. For example, both ENSO and NAO phenomena influence rainfall in Iberia (Rodo et al. 1997; Trigo et al. 2002). ENSO oscillation signals have been observed in the rainfall records of the past century in Iberia. There is clear evidence that the western part of the peninsula is more generally under the influence of NAO in winter while the eastern part is under the influence of ENSO in spring and autumn.

In: Nova Excerpts: Earth Sciences
Editor: Dominik Lucas

ISBN: 978-1-53619-038-0

Chapter 28

DEGLACIATION OF THE LATE WEICHSELIAN BARENTS SEA ICE SHEET*

Ivar O. Murdmaa and Elena V. Ivanova†
P. P. Shirshov Institute of Oceanology,
Russian Academy of Sciences, Moscow, Russia

ABSTRACT

According to the prevailing view, the Barents Sea shelf and surrounding archipelagoes were covered by a grounded ice sheet united with the Scandinavian ice sheet during the Last Glacial Maximum (LGM). Its decay started in the deep marginal troughs, the Bear Island Trough in the southwest, Kvitøya and Franz Victoria troughs in the north, ca. 19-18 cal ka BP with development of ice streams and related iceberg calving over the shelf break. Sediment cores and shallow boreholes recovered two distinct units of glaciomarine sediments. The lower Unit III is represented by a glaciomarine diamicton, a disordered mixture of coarse-grained iceberg rafted debris (from sand to pebbles and boulders), and very fine-grained clay (dominated by <0.001 mm fraction) released from the melting glacier ice. It overlies directly the eroded Mesozoic bedrock surface or non-

* The full version of this chapter is published *in Deglaciation: Processes, Causes and Consequences*, edited by Delbert Robbins, published by Nova Science Publishers, Inc., 2017.
† Corresponding Author Email: e_v_ivanova@ocean.ru.

marine basal moraine deposits (glacial till) in shelf depressions. The diamicton was deposited during the early deglaciation stage, which was apparently diachronous in the study area. The faunal and stable isotope data from the Kvitøya Trough supported by AMS-^{14}C dates suggest the older age of the early deglaciation stage in this narrow eastern Svalbard trough. Similar dates were previously published for the St. Anna Trough. However, the deglaciation of local marine-based ice sheet margins continues e.g., within the Franz Josef Land up to date with recent diamicton deposition. The diamicton is commonly overlain by Unit II with a sharp contact in between, deposited during the late deglaciation stage, when the iceberg calving and rafting ceased owing to glacier margins retreat to the land. The Unit II is mainly composed of fine-grained silty clay with thin fine-grained sand layers and lenses. It is often characterized by thin brown and gray lamination. Both laminated and lenticular sedimentary structures suggest pulsating deposition from glacier meltwater suspension flows likely accelerated by sea ice freezing brines cascading. Coarse IRD is rare or absent. Undulating seismic facies occur in the southern Barents Sea resembling sedimentary waves possibly indicating contourite origin of the thick Unit II. The upper boundary of the Unit II with overlying Holocene marine sediments appears to be also diachronous. Since the LGM until the Recent, the deglaciation was fostered by relatively warm surface and subsurface Atlantic water inflow into the Barents Sea through the western and northern deep troughs, respectively. The glacio-eustatic sea-level rise led to Atlantic water penetration beneath the glacier ice resulting in its melting from below thus promoting ice streams activity during early deglaciation. The distribution of iceberg rafted debris and its petrography depended on the provenances drained by ice streams as well as on the icebergs drift routes and their melting sites. Severe sea ice conditions were favorable for sub-ice sedimentation of the fine-grained diamicton matrix in small freshened basins between still glaciated shoals. Short down-slope suspension flows and seasonal brines cascading were main sedimentation mechanisms during the late deglaciation, although more extensive contour currents were possibly active in the southern Barents Sea. According to micropaleontological data, biological productivity was generally low during the deglaciation likely owing to restricted mixing of the subsurface saline Atlantic water and the surface water freshened by glacier melt-water discharge.

Keywords: deglaciation, lithostratigraphic units, microfossils, stable isotopes, Atlantic water

INTRODUCTION

The late Weichselian Barents Sea continental glacier represented the largest grounded marine based ice sheet of the last glacation in the Northern Hemisphere. According to the prevailing view (e.g., Svendsen et al. 2004), the Barents Sea shelf and surrounding archipelagoes were totally covered by an ice sheet united with the mainly land based Scandinavian ice sheet in the south during the Last Glacial Maximum (LGM) (Figure 1). Its boundaries followed the shelf brake in the west and the north. The eastern boundary in the Kara Sea is still uncertain and the maximum ice margin extent in the area was likely not synchronous with the timing of LGM in other margins (Hughes et al. 2016). The marine based ice sheet apparently covered the deep St. Anna Trough and possibly also the Voronin Trough. Further southward, the ice sheet boundary is commonly depicted along the north-eastern slope of the Eastern Novaya Zemlya trench through the Karskie Vorota Strait to the shallow Pechora Sea which emerged during the LGM sea level low-stand and was likely ice free. Behavior of the ice sheet margin in the south-easternmost Barents Sea – northern Pechora Sea during LGM and early deglaciation is still disputable among researchers based on different interpretation of detailed seismic survey, as well as shallow engineering borehole data (Polyak and Mikhailov 1996; Pavlidis and Polyakova 1997; Gataulin et. al. 2001; Krapivner 2006, 2009, and references therein). The boundary between Barents and Scandinavian ice sheets is arbitrarily drawn along the Nordcap – Kola trench. Both sheets had their own ice distribution centers which discharged glacier ice from both sides into the major ice stream flowing westward towards the Bear Island Trough that developed at the beginning of deglaciation (e.g., Siegert and Dowdeswell, 2002; Ottesen et al. 2005, 2008).

This chapter aims to highlight the influence of global and regional paleoceanographic changes on the deglaciation of the marine based Barents ice sheet since the last glacial maximum (LGM) until the onset of marine environment in the Holocene. A special attention is paid to the role of relatively warm Atlantic Water flow into the Barents Sea. However, it should be taken into account that even during the LGM, Atlantic Water seasonally penetrated into the Arctic Ocean via the Fram Strait and further eastward along the margin of Scandinavian-Barents ice sheet (e.g., Ivanova 2009; Xiao et al. 2015; Ivanova et al. 2016; Figure 1). Therefore, the relatively warm Atlantic water affected the configuration of the ice margin already at the onset of LGM and even much earlier, as shown in (Hughes et al. 2016).

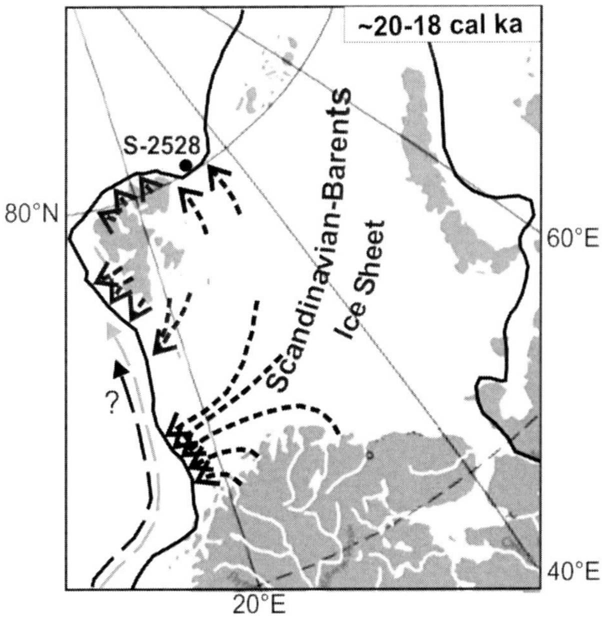

Figure 1. Scandinavian-Barents Sea Ice sheet extend and Atlantic Water flow along its margin at the end of Last Glacial Maximum (LGM), 20-18 cal ka BP (modified from Ivanova et al. 2016). Broken black and gray lines with arrows designate the inflow of surface and subsurface Atlantic Water, respectively. Ice streams are shown schematically by black dashed lines with arrows after (Landvik et al. 1998; Ottesen et al. 2005, 2008; Fransner et al. (in prep.)).

Here, we consider the deglaciation of the marine-based area of the Barents ice sheet. This process differs from the decay of land based ice sheets in dramatic influence of oceanographic parameters (temperature, salinity, stratification of water masses, their lateral and vertical dynamics, mixing with glacier melt water, global sea level change, etc.). These parameters can be reconstructed by proxy records preserved in glaciomarine sediment successions recovered by sediment cores and boreholes. Unfortunately, the scarcity of calcareous fossils (planktic and benthic foraminifers, mollusk shells) in the glaciomarine sediments of the Barents Sea strongly limits their dating and thus compilation of age models especially for the early deglaciation phase. Only a few datings are considered from the marine based Barents-Kara ice sheet in the comprehensive summary by Hughes et al. (2016). This led us to focus more attention on the lithostratigraphic approach in the present study, along with recent new radiocarbon datings and more adequate paleoceanographic interpretation of the data obtained.

In: Nova Excerpts: Earth Sciences
Editor: Dominik Lucas
ISBN: 978-1-53619-038-0

Chapter 29

INTRODUCTION: THE ANDES: GEOGRAPHY, DIVERSITY, AND SOCIOCULTURAL IMPACTS*

***Casey D. Allen*†**
Department of Geography and Environmental Sciences,
University of Colorado, Denver, CO, US

Stretching over 7200 km in a generally north-south direction along South America's western coast between 10 degrees north and 57 degrees south latitude – and traversing seven countries – the Andes Mountains represent a fascinating study locale. At their widest, the Andes span about 800km (roughly the distance from Denver, CO to Omaha, NE or the north-south length of Germany). They also represent the second highest mountain range (after the Himalyas) in the world, and host the highest elevation point in the Western Hemisphere (Mt. Aconcagua, 6962 meters).

From a physical geography persepctive, the Andes' influence on weather, climate, and landform processes remains strong. Generally acting as a barrier between the Pacific Ocean to the west and the rest of the mostly-

* The full version of this chapter is published in *The Andes: Geography, Diversity, and Sociocultural Impacts*, edited by Casey D. Allen, published by Nova Science Publishers, Inc., 2017.

† Address correspondence to: caseallen@gmail.com.

fertile continent to the east, resulting in a small rain shadow on the eastern slopes. In conjunction with the subtropical high pressure over the Pacific Ocean, the cool Humboldt/Peru current helps create a highly arid climate along the mountain's western slope, getting drier when moving closer to sea level. The result is the world's driest desert, the Atacama, where even bacteria have a difficult time living. High elevation mountains also mean glaciers, and the Andes contain nearly all of the world's tropical glaciers (i.e., glaciers located within the tropical latitudes).

Tectonics also plays a role in shaping the Andes. As the Nazca plate subducts under the South American plate, intense friction results, turning solid rock molten. The resultant magma tries to escape upward, creating volcanoes such as Galerasa and Nevado del Ruiz in Colombia, Cotopaxi and Sangay in Ecuador, several volcanic complexes in Peru and northern Chile, and Copahue, Llaima, and Villarica in Chile proper. Many Andean Volcanoes are located near highly populated areas and, alongside the sometimes powerful, tsunami-generating earthquakes experienced along South America's west coast, volcanoes remain one of the most impactful regional hazards.

In terms of human geography, the region's known occupational history stretches back to at least the Archaic period (~1800-9000 BC). Cultures specific to the Andes include the famous Inca and Nazca (who lived in the foothills of the Andes at times), and the lesser known but highly influential Chavín, Moche, Tiwanaku, Wari, Chachapoyas, and Chimú. These indigenous cultures often overlapped in both time and space, and as such, cultural diffusion occured readily. Ancestors of some of these ancient peoples still thrive in the same locations, practicing traditions that have been handed down over the centuries. Today, South America as a whole hosts a growing population of just over 420 million. Of that total, roughly 85 million live in the Andes and along the western coast. Still, the indigenous cultures continue to weave a vibrant tapestry of human occupance, just as their ancestors did, and this book hopes to add continuing threads to the loom. While a copendium on the Andes themselves would take up several volumes, this book focuses specifically on several recent studies in the Andes – from the southern Andes in Argentina to the central and northern

Andes of Peru, including the foothills – encompassing a wide array of geographic topics.

With that background in mind, following this short introduction, Gonzalez, Romero, and Garbarini begin the journey in the Argentinean patagonia section of the Andes, assessing internannual variablity of precipitation as tied to climate forcings related to several oscillations. Their findings have strong implications for not just other climate researchers, but for southern Andean residents as well. Then, in a related endeavor, Bisero and Gonzalez set their sights on modeling snow behavior in the Central Andes. Taking a regional approach allowed the researchers to better understand climate forcings related to snow accumulation, which could, in turn, have a strong impact on central Andean residents who rely on heavily on a limited supply of water.

Working at the Human-Environment Interface (HEI), Isla and Reindel use paleoclimate reconstruction to dissect migration and habitation patterns of ancient Andean peoples from the coast to the high elevations of southern Peru. Their comprehensive findings – taken from more than 1500 archeaological sites over the span of more than a decade – resulted in the first "fairly complete and detailed reconstructions" over a period of 10,000 years, revealing a more climate-related migratory and spatially-large trading monetary system than previously thought. In similar fashion, though not quite the aereal extent as Isla and Reindel's long-term project, Salazar focuses his HEI study in a narow basin on the Yungas called *Anfama* in the next chapter. Here, Salazar presents new evidence that challenges presuppostion of past inhabitants' landuse, and demonstrates instead a vibrant, dynamic group, "...of people who incorporated agriculture, pottery, and architecture into creating an original way of life." Salazar then expounds on what the findings might mean for the current residents of this often-overlooked, yet important region of the Andes.

The penultimate chapter utilizes evidence from two Tiwanaku-contemporary (~400-1000 AD) villages to dissect the notion of "differntial participation in interregional exchange" that supports "the emergence of inter-household socioeconomic differentiation." In this chapter, Beaule and DeLoge utilize macrofaunal camelid remains as a proxy for wealth to demonstrate sociodemographic changes between villages. Though useful in a historical context to better understand highland Bolivia's ancient social stratification, their research can also be applied to current villages as a way to assess on-going regional trade and subsistence practices. Focusing on more contemporary issues, readers are given an in-depth look at the recent "boom" of broom making in Peru's lower Andes in the concluding chapter. Here, Penn outlines a product origin disconnect among people in the region: although this recently resurrected cottage industry has spread across Peru – with thousands of people using the brooms daily – users are unaware of the brooms' origin. Penn's assessment remains bolstered by other pecuniary factors, such as the "War on Drugs" still being "fought" in the region (and throughout the Andes more generally), bringing to light important socioeconomic factors between cottage industry and product origin knowledge.

Edited volumes, I believe, represent an important contribution to our knowledge base. Rather than learning about a single person's findings (and often the author's own, unfettered opinions), collaborative efforts such as this book give a broader take on a subject, and especially so if the focus centers on a specific region. Even so, though the astute reader might notice a discernable pattern in the book's structure – the first two chapters outline mostly physical geography-related phenomena, the last two chapters center on more human geography-specific topics, and the middle chapters seem set at the human-environment interface (HEI) – as with any good (geography) book, each chapter has potential implications for the others, and the reader is admonished to look for and make those connections between the chapters. As others have noted, while the physical and human geography divide are often separated in vernacular, in reality and practice, the dividng line is not so straight as some might think, and can even be permeable (cf., Massey, 1999; Harrison et al., 2004; Allen and Lukinbeal, 2011; Inkpen and Wilson, 2013 – especially Chapter 10). Filled with up-to-date research by active researchers of the region, this book is no different. Hopefully the selection of topics piques the reader's interests in both content and region, perhaps even spurring-on the next generation of Andean researchers.

In: Nova Excerpts: Earth Sciences
Editor: Dominik Lucas
ISBN: 978-1-53619-038-0

Chapter 30

DROUGHTS AND FLOODS IN NORTHERN ARGENTINEAN PATAGONIA*

M. H. González[1,2,†], P. E. Romero[1] and E. M. Garbarini[1]
[1]Department of Athmospheric and Ocean Science,
FCEN, University of Buenos Aires, Buenos Aires, Argentina
[2]Sea and Atmospheric Research Center – CONICET/UBA,
UMI-IFAECI/CNRS, Buenos Aires, Argentina

ABSTRACT

The interannual variability of precipitation and river flow in the area of Comahue and northern Patagonia in Argentina is analyzed in this chapter. Both the monthly and annual precipitation and the mean annual flow data over the period 1985-2012 have been used to perform this study. For each station, the 25th and 75th percentiles of precipitation and flow were used to classify each year as dry (precipitation or flow less than 25th percentile), wet (greater than 75th percentile) or normal. Finally the years were classified as dry, wet or normal if the greater number of stations with in each basin belongs to that category.

* The full version of this chapter is published in *The Andes: Geography, Diversity, and Sociocultural Impacts*, edited by Casey D. Allen, published by Nova Science Publishers, Inc., 2017.

† Corresponding Author: gonzalez@cima.fcen.uba.ar.

Some parameters were defined and calculated in order to assess the intensity and the spatial coverage of all the extreme (wet and dry) events. Thus, all the extreme events are revealed in the different basins and their severity and affected areas in each case were documented.

Finally, the presence of significant climate forcing associated with extreme precipitation detected above was investigated. As the maximum signal was in spring, climate forcing in this season was analyzed in detail. Results indicate that the positive phase (negative) of the Indian Ocean Dipole, El Niño Modoki and the El Niño-Southern Oscillation is associated with the highest positive (negative) precipitation anomalies in spring in the Comahue region. Positive(negative) precipitation anomalies are also associated with the negative (positive) phases of the Antarctic Oscillation, the South Atlantic Ocean Dipole and the Multidecadal Atlantic Oscillation and with the weakening and eastward shift of the Pacific High and the intensification and shift to the south of the Atlantic High.

1. Introduction

The northern Argentinean Patagonia region covers the continental region of Argentina between 38° and 43°S, comprising the basins of the Limay, Neuquén and Negro rivers. The Andes Mountain extends west from north to south and eastwards the landscape is a quite dry plateau. The first two basins are relevant for hydropower generation and are located in an area called "Comahue" where rainfall prevails in winter and over the western high mountain. The other basins are arid but with valleys where growing fruit thrives. Eastern Patagonia precipitation is scarce and decreases towards the east with a maximum in summer. This socio-environmental complexity is expressed in processes of exploitation, inequitable distribution of natural resources, population movements and urban densification which includes the settlement in hazardous areas.

In order to understand the interannual variability of precipitation better some remote and regional climate forcings were analyzed. Some of them are related to sea surface temperature (SST) anomalies in remote oceans. The most known is the El Niño- Southern Oscillation (ENSO) which was proved to have a great influence on precipitation in South America(Ropelewsky and Halpert 1987, Vargas et al. 1999, Grimm et al. 2000, Vera et al. 2004, among others). The warm (cold) phase of ENSO is defined by a generalized heating (cooling) of the tropical Pacific Ocean. Ashok et al. (2007) defined special El Niño events called El Niño Modoki, where the heating is localized in the central tropical Pacific meanwhile a cooling is observed in eastern and western Pacific. They also defined an index called EMI which describes this type of Niños. The anomalies in the Indian ocean SST are related to the anomalies in precipitation in South America too. Saji et al. (1999) defined the Indian Ocean Dipole where the positive phase is associated with a heating of southwestern and a cooling of northeastern Indian Ocean. This effect is quantified through an index called OID (Ocean Indian Dipole).Chan et al. (2008) showed that the positive phase of OID is associated with increased precipitation in La Plata river Basin and decreased precipitation in central Brazil. These remote forcings act generating a Rossby wave train from the tropical oceans which propagates towards higher latitudes and displaces to the southeast over the Pacific Ocean arriving in southern Argentinean Patagonia. This pattern was well described by Kidson and Mo (2000) who defined the "Pacific South American Pattern." The Atlantic Ocean SST interannual variability also influences precipitation in South America. Therefore, the humidity advection from the Atlantic Ocean and from the Brazilian forest in northern South America is highly regulated by the Atlantic SST. The Atlantic Multidecadal Oscillation (AMO) is a variability SST mode in the North Atlantic Ocean and it was defined by Schlesinger (1994). It has a 60-80 year periodicity (VanOldenborgh et al. 2009) and it was related to rainfall anomalies in different areas (Enfield et al. 2001). Another dipole was defined in the South Atlantic ocean: the South Atlantic Dipole (SAODI) (Nnamchi et al. 2011), characterized by a positive phase when there is a heating in the northeastern (near the South African coast) and a cooling in the southwestern (near the Brazilian coast) parts of the South Atlantic ocean.

Some regional climate forcings influence precipitation anomalies, too. For example, the Southern Annular Mode which is quantified by an index called Antarctic Oscillation (AAO) and first defined by Thompson and Wallace (2000). Its positive phase is associated with low pressure anomalies over the South Pole and positive pressure anomalies in mid-latitudes, and therefore with an intensification of westerlies. This pattern prevents the energy exchange between latitudes and determines the frontal system trajectories.The AAO index was defined as the normalized pressure difference between 40°S and 70°S (Nan y Li 2003). Other regional factors, as the position and intensity of the South Atlantic (SAH) and the Pacific High (SPH), are also important to determine anomalies in precipitation in Argentina. A detailed description of all these factors can be found in Gonzalez et al. (2015, 2016).

The interannual rainfall variability in the northern Patagonia region was studied by Scarpati et al. (2014) and Gonzalez (2013) and some statistical prediction models were proved (Gonzalez y Herrera 2014, Gonzalez 2015). The main characteristics and the relation between river flow and precipitation were detailed in Romero and Gonzalez (2016).

The main objectives of this chapter are to describe a methodology to detect years with extreme precipitation anomalies generalized in the whole basins, to define some index to quantify the severity and the spatial coverage of the extreme events and to describe the remote and regional forcings that affect them.

In: Nova Excerpts: Earth Sciences
Editor: Dominik Lucas
ISBN: 978-1-53619-038-0

Chapter 31

STATISTICAL EVALUATION OF EXPERIMENTAL MODELS FOR THE PREDICTION OF SNOW IN THE CENTRAL ANDES*

N. K. Bisero[1,†] and M. H. González[2]
[1]National Weather Service, Buenos Aires, Argentina
[2]Department of Atmospheric Sciences FCEN/CIMA CONICET, University of Buenos Aires, Buenos Aires, Argentina

ABSTRACT

The aim of this chapter focuses on designing statistical models to predict the accumulation of snow in the region of the Central Andes in Argentina. This section describes previous research, the data set and the definition predictors which were used in the forecasting models. A Regional Average of Snow in the region of the Central Andes (RAS) was used for this research. The series RAS is representative of the precipitacion in the mountain because it was built using measurements in stations between 32° and 38°S. Stepwise and Cross-validation Method were also employed to build and prove the efficiency of the models.

* The full version of this chapter is published in *The Andes: Geography, Diversity, and Sociocultural Impacts*, edited by Casey D. Allen, published by Nova Science Publishers, Inc., 2017.

† E-mail: natykb@gmail.com.

Keywords: ENSO, Regression Analysis, Central Andes, snowpack, climate prediction

INTRODUCTION AND BACKGROUND

The Central Andes in Argentina are located between 32° and 38° south latitude, with an elevation of about 3500 m, decreasing from north to south with some higher peaks at 6000 m. To the east of the Andes Mountains, moisture comes from the north, in the summer months (December-January-February), season when precipitation is maximum. In this region's winter, snowfall mainly occurs in the high mountains. Still, rainfall remains scarce and, in terms of many related economic activities such as agriculture, the contribution of snowmelt to the river flows continues to play an important role for water supply in the region. The snowpack in this area represents the main source of water of most rivers in the region, and subsequent uses include domestic consumption, irrigation, industry, and electricity generation through dams (Masiokas et al. 2006). Given these important economic factors, it is important to better understand the climate variability in terms of predicting either precipitation or snow accumulation over the mountains, so all involved parties can plan appropriately.

It is well known that small variations on the Earth's surface, such as the sea surface temperature, may influence global atmospheric circulation as well as precipitation.

These regional oceanic heating or cooling patterns often act as remote forcing triggers and generate teleconnections. One of the most relevant patterns is the known El Niño-Southern Oscillation (ENSO), associated with anomalies of Sea Surface Temperature (SST) in the tropical Pacific Ocean that generate Rossby waves, which propagate meridionally from the tropics towards midlatitudes (Ropelewski and Halpert 1987; Kidson 1999; Nogues Paegle and Mo 2002). Compagnucci and Vargas (1998) observed that winters preceding a warm (cold) event are characterized by high (low) amount of snow in the mountains.

Many authors (e.g., Pittock, 1980; Quinn and Neal, 1983; Aceituno 1988; Rutlland and Fuenzalida 1991; Montecinos and Aceituno 2003) documented positive (negative) anomalies of winter precipitation in central Chile are related to warm (cold) ENSO phases. Other oceanic and atmospheric forcings influence precipitation as well. For example, the intensity of subtropical anticyclones and the belt of sub polar cyclones determine the intensity of the westerlies over the Pacific Ocean, generating a strong influence on the precipitation regime in Patagonia Argentina. Associated with these events, are the dynamics of the Antarctic Annular Mode (AAM), defined by Thompson and Wallace (2000).

Despite the limitations in seasonal forecasting (cf., Barnston et al. 2005; Quan et al. 2006; Kunmar 2006; Coelho et al. 2005), many research centers in the world have addressed the issue of improving both dynamic and statistical rainfall forecasts.

Dynamic models, in many cases, have limitations of spatial resolution, leading researchers to utilize statistical models when making such forecasts for limited areas. These statistical forecasts are usually based on the detection of the relationship between precipitation, circulation patterns and SST, which were observed previously to precipitation events.

In: Nova Excerpts: Earth Sciences
Editor: Dominik Lucas
ISBN: 978-1-53619-038-0

Chapter 32

PALPA AND LUCANAS: CULTURAL DEVELOPMENT UNDER CHANGING CLIMATIC CONDITIONS ON THE WESTERN SLOPE OF THE ANDES IN SOUTHERN PERU*

Johny Isla[1,†] and Markus Reindel[2]
[1]Centro de Investigación para la Arqueología y el Desarrollo (ANDES), Lima, Perú
[2]Deutsches Archäologisches Institut (DAI), Berlin, Germany
Kommission für Archäologie Außereuropäischer Kulturen (KAAK), Bonn, Germany

ABSTRACT

The Palpa valleys and its tributaries, in the coastal and highlands of southern Peru, are being investigated by a multidisciplinary team of

* The full version of this chapter is published in *The Andes: Geography, Diversity, and Sociocultural Impacts*, edited by Casey D. Allen, published by Nova Science Publishers, Inc., 2017.

† Email: isla.nasca@gmail.com.pe.

archaeologists and geographers from Peru and Germany, aiming to reconstruct the region's cultural processes in terms of a varied landscape and changing climate. Research has revealed a long cultural process that began more than 10,000 years ago and whose clearest evidence ranges from the early Archaic to Inca periods. During this research, more than 1500 archaeological sites have been recorded in surveys and excavations. Additionally, in cooperation with the Institute of Geography of the University of Heidelberg, palaeoclimatic changes that occurred during the Holocene were recorded in detail. Analyses of fluvial and aeolian sediments of the desert coast, as well as analyses of macroremains and pollen of drill cores from peat bogs of the highlands, demonstrate that climatic parameters changed gradually over time and influenced the living conditions and settlement patterns of the pre-Hispanic societies.

Keywords: Palpa Valley, Peru, paleoclimate reconstruction, pre-Hispanic, highlands

INTRODUCTION

During pre-Hispanic times, the Andes' inhabitants adapted to specific conditions over a variety of ecological zones, where they developed lifestyles and corresponding economic activities to each natural environment. Evidence of these traditional regimes can be seen in the remains of their settlements and the material culture of each region. Even though archaeologists have the challenge of rebuilding a variety of lifestyles and cultural patterns in the midst of a mosaic of habitats, today evidence of adaptive livelihoods can still be identified and studied with archaeological methods. Still, the reconstruction of living conditions in ancient times remains complicated when considering unstable weather conditions over time. Evidence gained through systematic study of settlements and associated materials in a fairly large area, points to changes in temperature and humidity conditions significantly affecting ecological zone boundaries, causing changes in the economic systems of the societies established in them. This, at the same time, resulted in the displacement of populations and changing settlement patterns.

The constant movement of people and exchange of resources and its derivatives between different ecological zones has been considered an ancient Andean tradition, described with the help of ethnohistorical sources and development models as vertical control of a maximum of ecological zones (Murra 1975). In many cases, these remain tied to both settlement patterns and economic pre-Hispanic systems that reflect such verticality. To date, however, any archaeological project that has investigated these economic and population dynamics of a macro-region, has not necessarily taken into account the Andes' ecological diversity in terms of climate.

From this perspective, in the framework of the archaeological projects "Nasca-Palpa" and "Palpa-Lucanas," directed by the authors, we consider the study of the cultural development and economic systems of the pre-Hispanic populations established in the Palpa valleys and their headwaters, which comprise various ecological zones on the western slope of the Andes.[1] One of the main objectives of the study was the analysis of patterns of regional settlement with archaeological methods covering all known periods, in conjunction with the development of climate and environmental changes throughout the research area and adjacent points.[2] Finally, we have analyzed the relations that existed between culture and climate, focusing on their implications in the long-term cultural processes in the region.

[1] These studies were developed in the framework of a research project called Transecta Andina, which was completed between 2002 and 2011, with funding from the Federal Ministry of Education and Research of Germany. Initial investigations in relation with the geoglyphs were sponsored by the Swiss-Liechtenstein Foundation for Archaeological Research Abroad (FLSA).

[2] Studies on climate and the environment were conducted in cooperation with geographers from the University of Heidelberg and Cologne, specialists in paleogenetics of the University of Goettingen and geologists specialized in the analysis of isotopes of the Collection for Geology and Paleontology of the State of Bavaria, Münich.

In: Nova Excerpts: Earth Sciences
Editor: Dominik Lucas
ISBN: 978-1-53619-038-0

Chapter 33

THE YUNGAS OF THE SOUTH ANDES AND THEIR KEY ROLE FOR THE ONSET OF EARLY PRE-COLUMBIAN VILLAGES*

***Julián Salazar*†**
CEH Segreti (CONICET) and Universidad Nacional de Córdoba,
Córdoba, Argentina

ABSTRACT

Environmental characteristics of the Andean Yungas have been traditionally interpreted as constraining factors limiting cultural development based on intensive agropastoral economies. This idea was the milestone for an aprioristic view that dismissed the relevance of the Andean eastern slopes throughout the region's human history. That narrative, however, lacked of strong data support and reproduced old deterministic suppositions. This chapter presents new archaeological data coming from Anfama, a narrow basin located in the montane forest (the higher altitudinal ecozone of the Yungas), in order to demostrate the key role this region played on the development and consolidation of early

---

* The full version of this chapter is published in *The Andes: Geography, Diversity, and Sociocultural Impacts*, edited by Casey D. Allen, published by Nova Science Publishers, Inc., 2017.

† Email: juliansalazar@ffyh.un.edu.ar.

villages. Speculation about what these new findings mean in terms of current local populations are also discussed.

Keywords: Andes, Yungas, archaeology, early villages

INTRODUCTION

The southern piedmont of Cumbres Calchaquíes (NW Tucumán province, Argentina Republic) constitutes a wild and almost inaccessible area of the Andes because of the dense forest vegetation, the copious seasonal rainfalls, and the abrupt and steep terrain. These environmental characteristics have set the basis for an aprioristic supposition that dismissed the relevance of the Andean eastern slopes in the human history of the region, based on a lack of strong evidence, that subsequently (re)produced deterministic hypotheses.

As a result of archaeological research carried out over the last two years in the Anfama river basin, our team recorded a strong human occupation dating back to 2300 years BP. This apparently hard landscape was the homeland for development of several cultural processes that involved dense occupied settlements with large and solid stone houses, high altitude adapted agriculture systems, original pottery styles, and a complex variety of material devices for ancestral worship. Among the last, the *huancas* stand out: beautifully carved stone sculptures representing humans, animals, and mythic entities. These represented the main pieces of native Pre-Columbian art scattered across the landscape. This local development was part of the overall formation processes – as well as the growth and abandonment of early villages in the south Andes during the first Millennium AD – rather than a marginal phenomenon alongside from highland "nuclear" developments.

In this chapter, following an overview of the upper Yungas' ecozone (montane forest) and landscape characteristics, I analyze previous narratives about the relationship between southern Andean Yungas geographical characteristics and human social evolution. Finally, I present the archaeological evidences of early village development in the area, outlining the range (and role) of geographical conditions in the evolution of this process, while also critically analyzing the consequences of archaeological narratives for current inhabitants of the area.

In: Nova Excerpts: Earth Sciences
Editor: Dominik Lucas
ISBN: 978-1-53619-038-0

Chapter 34

Trading Up: Interregional Exchange and Inter-Household Wealth Differentiation in Two Highland Bolivian Villages†

***Christine D. Beaule*[1,*] *and Alana N. DeLoge*[2]**
[1]University of Hawai'i at Mānoa, Honolulu, HI, US
[2]University of Pittsburgh, Pittsburgh, PA, US

Abstract

This chapter tests the model that differential participation in interregional exchange underwrites the emergence of inter-household socioeconomic differentiation, using archaeological assemblages from Jachakala and Condor Chinoka, two Tiwanaku-contemporary (ca. AD 400-1000) villages in highland Bolivia. Analyses of trade goods, including artifacts imported from or in the style of ceramics from the Tiwanaku State, reveal significant internal variation in households' trade and consumption of exotics. We predict greater inter-household wealth differentiation at

† The full version of this chapter is published in *The Andes: Geography, Diversity, and Sociocultural Impacts*, edited by Casey D. Allen, published by Nova Science Publishers, Inc., 2017.

Corresponding Author: beaule@hawaii.edu.

Condor Chinoka than Jachakala, because its ceramics include more imported vessels, whereas Jachakala has more local versions. Therefore, Condor Chinoka's greater participation in trade networks directly linked it to Tiwanaku and the eastern Bolivian lowlands, giving households greater opportunities to accumulate wealth and prestige. We use camelid faunal remains as a proxy for wealth at both sites, and find significant differences at Jachakala in households' access to the most valued cuts of meat in spatial patterns correlating with differences in imports. However, early wealth differences predate contact with Tiwanaku (and corresponding increases in exchange). Condor Chinoka exhibited no significant socioeconomic variability. We argue that Jachakala households intensified subsistence and craft production to meet other goals related to its intensified and variable cross-cultural interactions across the southern Andes. At Condor Chinoka, some households traded more than their neighbors, but they were unable to turn differential access to fancy pottery into wealth. Finally, we explore the broader implications of these findings for discussions about the relationships between trade, cross-cultural contact, and the development of hierarchy in small village communities.

Keywords: household archaeology, Bolivia, Tiwanaku, Andes, interregional exchange, faunal remains, complex society

Introduction and Site Setting

Throughout prehistory and history, households and communities have had to work out ways to adapt to local natural and social environmental conditions in their efforts to provision themselves with the goods and services that they need. Because few households can produce everything on their own, this has required a wide range of acquisition strategies, ranging from participation in formal marketplaces servicing defined regions (or spanning the globe) to direct reciprocal barter between producers, with or without intermediary merchants. Thus the scale and mechanisms of exchange can vary enormously. Formal systems of trade networks, merchants, and marketplaces are known to have functioned in most ancient state systems (Hirth and Pillsbury 2013), but not in the Andean region of South America before the Spanish conquest. Despite the development of a number of prehistoric states and empires, Andean households and communities acquired long-distance goods through well-established trade networks of merchants and camelid (llama and alpaca) caravans that facilitated the redistribution of both everyday goods (e.g., food, pottery, stone tools) and luxury goods (e.g., decorated religious objects, semi-precious stones, rare species of marine fish and shell, and feathers from Amazonian birds). In part because these networks predate state-level polities in the Andes by millennia (Dillehay 2013), and in part because they functioned without or outside of state institutions, they are a critical tool in the economic systems of indigenous Andean populations' adaptation to this challenging region. The Andes are a young group of mountains, tectonically active, remarkably difficult to traverse at some points, arid, cold, and unforgiving. The peoples who have lived on and around them for millennia have devised adaptations to significant variability in temperature and precipitation by developing varieties of food crops like the potato to local conditions. The small number of domesticated animals available to prehistoric Andean peoples - llamas and alpacas (exploited as beasts of burden, for meat, wool, tendons, bone and blood), the dog and guinea pig – meant too that products from the Pacific coast to the jungle were distributed through traders who travelled on foot. Trade networks, and the distributions of goods over space that are their primary material signature, are thus one important component to the study of how Andean peoples adapted to the unique opportunities and obstacles presented by their environment.

Trade and exchange are reconstructed archaeologically by the distribution of goods that came from elsewhere, such as marine shell found at a site far inland. They are, as Hirth and Pillsbury (2013, 6) write, "forms of dyadic distribution between mutually interacting parties." Trade can be defined as "a form of balanced exchange that may occur within marketplaces and through social partnerships." Exchange is conceptually broader, signifying the movement of not only physical goods, but also the transfer of ideas, material styles, and people themselves. The exchange of goods "can encompass transfers that are balanced or unbalanced, immediate or delayed, and confined to socially established partners or open to all" (2013, 6). The mechanisms of trade are often most visible in the context of ancient states; most, but not all, complex societies in the ancient world had some form of marketplaces. Despite the tendency to focus archaeological studies of trade systems on the needs and activities of state-level institutions and elite activities, the far more numerous commoner households intersected with trade networks and marketplaces in highly variable ways and levels of intensity (2013, 7). Marketplaces were clearly sites where trade and exchange relationships were formally constituted, but evidence for marketplaces in the central Andes is quite scarce. Hitherto available evidence overwhelmingly supports the characterization of central Andean economies as noncommercial, largely lacking both professional merchants and formal marketplaces (2013, 11), although this longstanding conclusion is being challenged by some recent research (e.g., Mayer 2002, 2013). Instead, central and southern Andean state-level economies were "generally based on elaborate redistributive, reciprocal, and administered trade systems" (Stanish 2010, 187). Households produced and consumed much of their own subsistence and craft goods, supplemented with goods, ideas, material styles, and perhaps foreign producers through local fairs. These networks connecting households with goods and individuals from distant places were well established before state-level polities and institutions emerged in the Andes. We agree with Stanish (2010, 187) that "Long-distance interregional exchange of many kinds of items was also robust and historically deep."

In the southern Andes, trade caravans were an important decentralized mechanism of trade from very early in prehistory. Several scholars have recently emphasized the importance of understanding that "Domestic provisioning relies on the operation of multiple forms of resource acquisition in weakly integrated market systems…that included household-to-household reciprocal exchange, barter, limited market exchange, and redistribution based on verticality" (Hirth and Pillsbury 2013, 12). Dillehay (2013) adds cyclical residency and household-to-household reciprocity to that list of mechanisms. Goldstein's work (2005, 2013) finds evidence for zonal complementarity in southern Peru's Moquegua Valley; this model (also called verticality) describes direct exploitation of products from different ecological zones by groups of households linked by *ayllu* (kinship group) membership in direct or indirect, reciprocal and decentralized exchanges. Nielsen's (2013) "intermodal approach" examines goods moving along communication corridors between transport sites along trade routes from coast to highlands. Goods such as ceramics, lithics, lapidary goods, salt, feathers, ash, fruit, eggs, and game moved through caravan trade via reciprocal barter and open and decentralized circulation networks (even under the Inka, and usually outside of state or elite control). This approach moves the locus of analysis from centralized marketplaces to the households, villages, towns, and other sites where goods were consumed, purchased or exchanged, produced, and discarded. This decentralized approach to studying trade and exchange allows the simultaneous consideration of local and long-distance trade goods from several perspectives: as markers of status, wealth, part of households' domestic economy, and where relevant, as indicators of unequal socioeconomic relationships between parties in those exchanges.

However, research on trade and exchange has a second, often neglected implication. It is important to understand distributions of goods not only in terms of household provisioning strategies (explanations of how households and communities fulfilled their physical and social needs), but also to understand what impact those (distributions of) goods had on social and political relationships within and between households. The two case studies presented in this chapter provide a valuable opportunity to examine the distributions *and* impacts of the subsistence and prestige goods produced or traded by two small prehistoric village communities, and to contextualize those activities' impacts on their changing relationships with each other. Between ca. A.D. 600-1000, the ancient Tiwanaku State emerged in the northern Bolivian highlands near Lake Titicaca as a powerful economic and ceremonial center that impacted a large area including the southern Peruvian and northern Chilean coast, central Bolivian Altiplano, and into the eastern Bolivian lowlands. Tiwanaku trade goods found throughout these areas include decorated ceramic vessels used in religious rituals, specific tool types (e.g., camelid mandible tools), musical instruments, ceremonial hallucinogenic drug ingestion paraphernalia (snuff trays and tubes), and perhaps subsistence goods such as maize and freeze-dried meat (*ch'arki*). Artifacts from Tiwanaku, the Pacific coast, the Peruvian highlands, and

Bolivian lowlands were recovered from household contexts at the two prehistoric sites of Jachakala and Condor Chinoka (Figure 1). Neither village community yielded evidence for a political elite or formally institutionalized leadership. Both were by all indications small, relatively egalitarian villages of farmers and herders who met their subsistence needs through herding, hunting, limited fishing, and direct farming. Households in both communities also participated in long-distance exchange networks that brought semi-precious stones (obsidian and ópalo), basalt stone tool blanks, painted ceramics, and marine shell from distant locations, and in some cases, substantial quantities throughout their histories.

One way that archaeologists sometimes tackle the analysis and interpretation of distributions of trade goods is to tie differences through time in their appearance, and differential distributions within a given time period, to other changes at those sites. In the case of Jachakala, significant changes in the community in the kinds and relative weight accorded to household-level subsistence activities (farming and herding) were traced through the distributions of the material remains of those activities. The import of stone tool blanks, production of household tools (primarily including agricultural hoes, but also scrapers, projectile points, knives, and other everyday tools) and herding (measured indirectly through the distributions of camelid bones from animal butchering and consumption), and utilitarian (unpainted) pottery were mapped through time and across the village. Approximately halfway through the site's history, Tiwanaku-style ceramics and other imported artifacts appeared; in some cases, their distribution was limited to one residential area but not the other, while in other cases, all households had access to certain kinds of imported goods. The appearance and differential distribution of Tiwanaku trade goods coincided with the differential or exclusive consumption of trade goods (marine shell, obsidian, etc.) from elsewhere. As wealth and status differences between Jachakala's households emerged over the centuries, the role of trade goods in the manifestation of those sociopolitical shifts also changed. We have argued in the past (Beaule 2002, 2012) that these patterns of trade goods bolster arguments that Jachakala manifested early complexity (social, political, and/or economic inequalities), but that the emergence of those intra-community wealth differences depended on access to valued food (e.g., certain cuts of camelid meat) rather than participation in trade networks themselves, which all seem to have done without exception. Focusing on trade goods as markers of status, and as indicators of differential participation in household- or village-wide ritual activities, was a more productive approach than equating trade goods alone with wealth.

The Condor Chinoka case, on the other hand, complicates this model correlating the origins of complex society with differential participation in trade networks. This site is another one with occupations pre-dating and contemporary with the expansion of the Tiwanaku State. Condor Chinoka's artifact assemblages, subjected to the same range of analyses as the Jachakala project, yielded no substantial evidence for inter-household wealth or status differences during the community's history (DeLoge and Beaule 2006). However, the site had much higher proportions and densities of trade goods, particularly ritual ceramic vessels, from Tiwanaku or from other Tiwanaku-contemporary sites in the Bolivian eastern lowlands. Condor Chinoka's strategic location at the end of the increasingly narrow Obrajes valley on the eastern edge of the Bolivian altiplano means that camelid caravans passing down the valley's length had to have passed within a stone's throw of the village. It was thus located at a place on the landscape where it was perhaps far likelier than Jachakala, sited on a very wide, mostly flat highland plain, to have regular interaction with caravans of traders crossing the Bolivian highlands, linking it to Tiwanaku, the Pacific coast, the eastern lowlands (where coca, maize, and fruit grew in abundance), and the Amazonian basin beyond that. Why then did long-distance trade goods not spark a similar internal development of inter-household wealth differences, as it had at Jachakala?

This chapter presents a summary of the analyses and distribution patterns of trade goods at these two sites, and uses a comparative perspective to explore a set of related questions about archaeological uses of trade goods as measures of social and subsistence provisioning, social status, political prestige, religious ideology, and ritual practices.

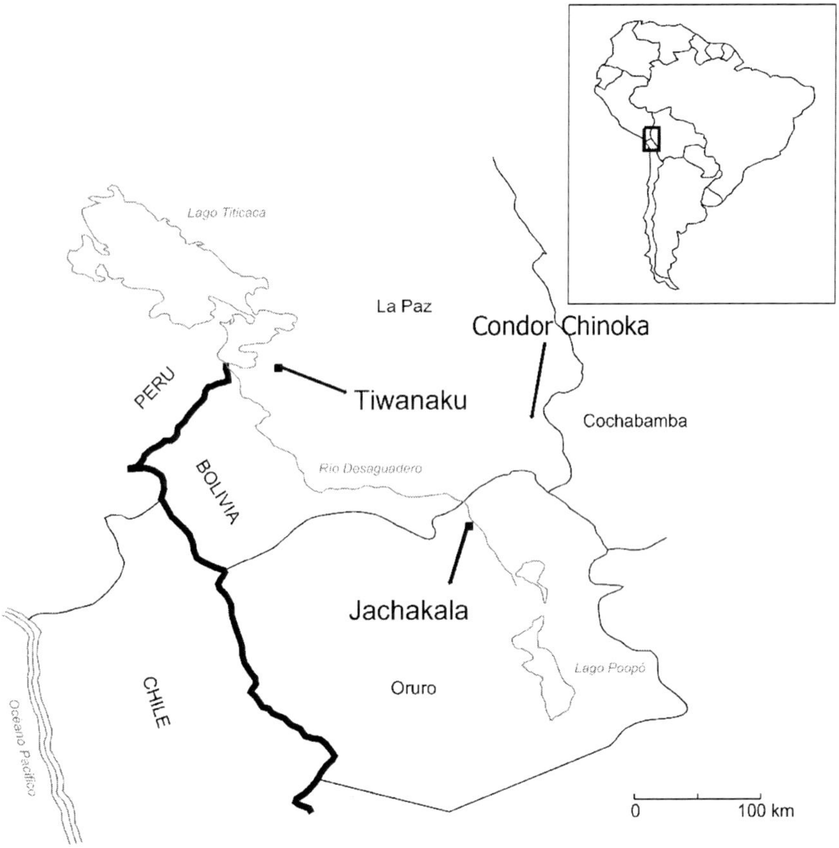

Figure 1. Map showing locations of Jachakala, Condor Chinoka, and Tiwanaku in Bolivian highlands.

In: Nova Excerpts: Earth Sciences
Editor: Dominik Lucas
ISBN: 978-1-53619-038-0

Chapter 35

PIASSABA BROOM PRODUCTION AND FIBER HARVEST AMID CHALLENGING CONDITIONS IN THE LOWER ANDES: A CASE FROM PERU*

James W. Penn, Jr.†
Grand Valley State University, Allendale, MI, US

ABSTRACT

This study examines how many socioeconomic factors (including the "War on Drugs") combine to explain the "boom" in broom production made from fibers of the piassaba palm (*Aphandra natalia*) in the Huallaga River valley of the lower Andes during the last 20 years, and how this resurrected cottage industry has spread from the Andes to towns and cities across Peru. Despite widespread poverty and low incomes generated from fiber collection and making the brooms, along with the limited distribution and scarcity of the palms in the wild, the industry has become an important source of income for large numbers of both rural and urban families in

* The full version of this chapter is published in *The Andes: Geography, Diversity, and Sociocultural Impacts*, edited by Casey D. Allen, published by Nova Science Publishers, Inc., 2017.

† Email: pennji@gvsu.edu

Peru. Meanwhile, Peruvians all across the country utilize the brooms, with almost no knowledge of their origin.

Keywords: broom, lower Andes, piassaba, Huallaga River, Peru, cottage industry

INTRODUCTION

Numerous studies in the last few decades have examined the use, economic importance and sustainability of non-timber forest products (NTFPs) extracted from the Amazon (Posey and Balée 1989, Butler 1992, Nepstad and Schwartzman 1992, Neumann and Hirsch 2000). Fruits, thatch, fuel, fibers, medicines, dye, and myriad other products come from a multitude of Amazonian forest species. The economic potential of NTFPs in Amazonia remains enormous and underappreciated, and we are only beginning to understand their economic value (Smith 1999). For example, NTFPs from palm trees have long been known as sources of subsistence and Amazonian commerce, while supplying large national and international markets. By the early 1990s palm heart harvested from the *açai* palm (*Euterpe oleracea*) in eastern lowland Brazil was providing almost 30,000 jobs and annual revenues of US $300,000 million (Clay 1997). In the 21st century, this industry, using the same palm species, had expanded into a global açai fruit business worth more than US $1 billion (Brondizio 2008).

In the more isolated and less populated western Amazon, especially important markets for NTFPs extracted from the wild have developed in Peru (Padoch 1988, Vasquez and Gentry 1989, Peters and Hammond 1990, Coomes 1995), with palm species from the lowland forests as some of the most important and abundant sources of these products in the Peruvian Amazon (Peters et al. 1989, Padoch 1992, Penn 2008, Brokamp et al. 2011, Smith 2015). However, much less attention has been given to NTFPs that come from the montane forests or *selva alta* of the lower Andes of the Amazon basin. This may have to do with the fact that even the more recent studies that address nature, culture and conservation in the Andes (e.g., Gade 1999, Brown and Mitchell 2000, Carey 2010, Borsdorf and Stadel 2015) still tend to focus mostly on the high Andes where forest cover is rare or absent, or the jungle margins, with scant mention of the foothills, or selva alta.

The selva alta refers to the foothills of the eastern Peruvian Andes, which range from 500 to 2000m in elevation (Young and León 2000). Also called *rupa-rupa* in Quechua, the selva alta lies between the *yunga*, which is the narrow transition zone that drops sharply off the Andean highlands to the start of the forested zone (**Borsdorf and Stadel 2015) and the *selva baja*, which covers most of the Amazon basin at altitudes less than 500m.** With the exception of *uña de gato* (*Uncaria tomentosa*), a liana widely marketed from the selva alta of Peru for its medicinal properties (de Jonge et al. 1999), few studies have examined NTFPs from this biodiverse altitudinal zone. It is in these threatened forests where the piassaba palm (*Aphandra natalia*) is found in the Andes.

Piassaba ("piasaba," "piazaba") is a very useful and heavily exploited palm in the montane forests of the eastern slopes of the Andes. Called *virote huasi* in the Quechua dialect of San Martín province of Peru (Mendoza 2013), the fibers from the palm are used to make enormous quantities of brooms that are used throughout the country. This study examines the environmental and socio-economic setting in relation to this NTFP of Peru, and the cultural changes impacting the people who work in the wild harvest and trade of piassaba fibers, and the broom making industry and trade. This palm resource has had a wide impact on both rural and urban society of Peru, amid very challenging political pressures and cultural conditions.

INDEX

#

A

D

E

F

G

K

L

M

N

O

P

S

T